T0296213

Cambridge Tracts in Mathematics
and Mathematical Physics

GENERAL EDITORS
G. H. HARDY, M.A., F.R.S.
E. CUNNINGHAM, M.A.

No. 10
An Introduction to
the study of Integral Equations

AN INTRODUCTION TO THE
STUDY OF INTEGRAL EQUATIONS

BY

MAXIME BÔCHER, B.A., Ph.D.

Professor of Mathematics
in Harvard University

CAMBRIDGE
AT THE UNIVERSITY PRESS
1926

CAMBRIDGE
UNIVERSITY PRESS

University Printing House, Cambridge CB2 8BS, United Kingdom

Cambridge University Press is part of the University of Cambridge.

It furthers the University's mission by disseminating knowledge in the pursuit of education, learning and research at the highest international levels of excellence.

www.cambridge.org
Information on this title: www.cambridge.org/9781107493490

© Cambridge University Press 1926

First edition 1909
Second edition 1914
First published 1926
Reprinted 1926
Re-issued 2015

A catalogue record for this publication is available from the British Library

ISBN 978-1-107-49349-0 Paperback

PREFACE

IN this tract I have tried to present the main portions of the theory of integral equations in a readable and, at the same time, accurate form, following roughly the lines of historical development. I hope that it will be found to furnish the careful student with a firm foundation which will serve adequately as a point of departure for further work in this subject and its applications. At the same time it is believed that the legitimate demands of the more superficial reader, who seeks results rather than proofs, will be satisfied by the precise statement of these results as italicised, and therefore easily recognized, theorems. The index has been added to facilitate the use of the booklet as a work of reference.

In these days of rapidly multiplying voluminous treatises, I hope that the brevity of this treatment may prove attractive in spite of the lack of exhaustiveness which such brevity necessarily entails if the treatment, so far as it goes, is to be adequate.

I wish to thank Professor Max Mason of the University of Wisconsin who has helped me with some valuable criticisms; and I shall be grateful to any readers who may point out to me such errors as still remain.

MAXIME BÔCHER.

Harvard University,
Cambridge, Mass.
November, 1908.

This second edition is a reprint of the first, in which, however, such errors as have come to my notice have been corrected. Of these the most serious (on pages 17 and 62–64) were called to my attention by Professor D. R. Curtiss and Dr W. A. Hurwitz respectively.

M. B.

December, 1913.

CONTENTS

AN INTRODUCTION TO THE STUDY
OF INTEGRAL EQUATIONS

Introduction. The theory and applications of integral equations,
or, as it is often called, of the inversion of definite integrals, have come
suddenly into prominence and have held during the last half dozen
years a central place in the attention of mathematicians. By an integral
equation* is understood an equation in which the unknown function
occurs under one or more signs of definite integration. Mathematicians
have so far devoted their attention mainly to two peculiarly simple
types of integral equations,—the linear equations of the first and second
kinds,—and we shall not in this tract attempt to go beyond these cases.
We shall also restrict ourselves to equations in which only simple (as
distinguished from multiple) integrals occur. This restriction, however,
is quite an unessential one made solely to avoid unprofitable complications at the start, since the results we shall obtain usually admit of an
obvious extension to the case of multiple integrals without the introduction of any new difficulties †. In this respect integral equations are
in striking contrast to the closely related differential equations, where
the passage from ordinary to partial differential equations is attended
with very serious complications.

The theory of integral equations may be regarded as dating back at
least as far as the discovery by Fourier of the theorem concerning
integrals which bears his name ; for, though this was not the point of
view of Fourier, this theorem may be regarded as a statement of the
solution of a certain integral equation of the first kind‡. Abel and
Liouville, however, and after them others began the treatment of
special integral equations in a perfectly conscious way, and many of
them perceived clearly what an important place the theory was destined
to fill§.

---

* The term Integral Equation was suggested by du Bois-Reymond. Cf. *Crelle*,
vol. 103 (1888), p. 228.

† Another extension, in which serious complications do not usually arise, is to
systems of integral equations. We do not consider such systems in this Tract.

‡ Cf. the closing page of this Tract.

§ Cf., besides the article of du Bois-Reymond already cited, some remarks by
Rouché, Paris *C. R.* vol. 51 (1860), p. 126.

As we shall not, except in one relatively unimportant case, take up any of the applications of the subject, it may be well to say explicitly that like so many other branches of analysis the theory was called into being by specific problems in mechanics and mathematical physics. This was true not merely in the early days of Abel and Liouville, but also more recently in the cases of Volterra and Fredholm. Such applications of the theory, together with its relations to other branches of analysis *, are what give the subject its great importance.

1. Some Preliminary Propositions and Definitions. In order to avoid interruptions in later sections, we collect here certain propositions of the integral calculus for future reference.

We shall have to deal with functions of one and of two variables. The independent variables, which we will for the present denote by x and (x, y) respectively, are in all cases real. In fact, in order to avoid unnecessary complications we will assume that, unless the contrary is explicitly stated, *all quantities we have to deal with are real.*

The range of values of the single argument x is usually

$$I \qquad\qquad a \leqq x \leqq b.$$

We shall speak of this in future simply as the interval I.

In the case of functions of two variables, two cases have to be considered. Interpreting (x, y) as rectangular coordinates in a plane, we sometimes consider the square

$$S \qquad\qquad \begin{cases} a \leqq x \leqq b \\ a \leqq y \leqq b \end{cases},$$

and sometimes the triangle

$$T \qquad\qquad a \leqq y \leqq x \leqq b.$$

It should be noticed that the three regions we have just defined, I, S, T, are closed regions, that is they include the points of their boundaries.

In order to avoid long circumlocutions we lay down the following

DEFINITION. *We say that the discontinuities of a function of (x, y) are regularly distributed in S or in T if they all lie on a finite number of curves with continuously turning tangents no one of which is met by a line parallel to the axis of x or of y in more than a finite number of points.*

In order to make the enunciation of some of our results simpler, we will assume once for all that the functions we deal with are defined even

* Cf., for instance, much of Hilbert's work.

at the points of discontinuity, at least in the cases where they remain finite in the neighbourhood of such points.

The following theorem will be important for us. We state it first for the case of the region S.

THEOREM 1. *If the two functions $\phi (x, y)$ and $\psi (x, y)$ are finite in S and their discontinuities, if they have any, are regularly distributed, the function*

$$F(x, y) = \int_a^b \phi (x, \xi) \psi (\xi, y) d\xi$$

is continuous throughout S.

The truth of this theorem becomes evident if we interpret (x, y, ξ) as rectangular coordinates in space. It is then clear that the function under the integral sign is finite throughout the cube

$$a \leqq x \leqq b, \quad a \leqq y \leqq b, \quad a \leqq \xi \leqq b,$$

and becomes discontinuous in this cube only at points on two sets of cylinders whose generators are parallel respectively to the axes of x and y. Moreover these cylinders are so shaped that any line $x = x_0$, $y = y_0$ in this cube meets them at only a finite number of points.—The formal proof, based on these or similar considerations, presents no difficulty, and we leave it for the reader.

COROLLARY. *If $\phi (x, y)$ and $\psi (x, y)$ are finite in T and their discontinuities, if they have any, are regularly distributed, the function*

$$H(x, y) = \int_y^x \phi (x, \xi) \psi (\xi, y) d\xi$$

is continuous throughout T.

This is merely a special case of Theorem 1. For if we define ϕ and ψ to have the value zero everywhere outside of T, it is clear that they satisfy the conditions of Theorem 1 throughout S and that the function $F (x, y)$ reduces to $H (x, y)$.

If $\phi (x, y)$ satisfies the conditions of Theorem 1, the double integral of ϕ extended over S may be evaluated in either one or two ways as an iterated integral* and we thus get the formula

$$\int_a^b \int_a^b \phi (x, y) \, dy \, dx = \int_a^b \int_a^b \phi (x, y) \, dx \, dy.$$

If, in particular, ϕ vanishes everywhere outside of T, we get

---

* By a double integral we understand the limit of a sum obtained by dividing up the region in question into pieces both of whose dimensions are small. By an iterated integral, the integral of an integral.

DIRICHLET'S FORMULA*. *If φ is finite in T and its discontinuities, if it has any, are regularly distributed, then*

$$\int_a^b \int_a^x \phi(x, y)\, dy\, dx = \int_a^b \int_y^b \phi(x, y)\, dx\, dy.$$

This formula admits of extension to certain cases in which the integrand does not remain finite in T. The most general case of this sort which we shall have occasion to use is contained in the following statement, for a simple proof of which we refer to the first part of a paper by W. A. Hurwitz†:

DIRICHLET'S EXTENDED FORMULA. *If φ (x, y) is finite in T and its discontinuities, if it has any, are regularly distributed, and if* λ, μ, ν *are constants such that*

$$0 \leqq \lambda < 1, \quad 0 \leqq \mu < 1, \quad 0 \leqq \nu < 1,$$

then

$$\int_a^b \int_a^x \frac{\phi(x, y)\, dy\, dx}{(x-y)^\lambda (b-x)^\mu (y-a)^\nu} = \int_a^b \int_y^b \frac{\phi(x, y)\, dx\, dy}{(x-y)^\lambda (b-x)^\mu (y-a)^\nu}.$$

Finally we turn to some theorems concerning functions of a single variable.

THEOREM 2. *If φ (x) is finite and has only a finite number of discontinuities in I, the function*

$$\Phi(x) = \int_a^x \frac{\phi(\xi)\, d\xi}{(x-\xi)^\lambda} \qquad (\lambda < 1)$$

is continuous throughout I, including the point a, where it vanishes‡.

To prove this we introduce the new variable of integration

$$s = \frac{\xi - a}{x - a}.$$

Then

$$\Phi(x) = (x-a)^{1-\lambda} \int_0^1 \frac{\phi[a + s(x-a)]}{(1-s)^\lambda}\, ds = (x-a)^{1-\lambda} \Psi(x)$$

$$(a < x \leqq b).$$

---

* Cf. *Crelle's Journal*, vol. 17 (1837), p. 45.

† *Annals of Mathematics*, vol. 9 (1908), p. 183. This result may also be deduced from a general theorem of de la Vallée Poussin. Cf. the *Cours d'Analyse* of this author, vol. 2, pp. 89—95.

‡ We *define* the symbol $\int_a^a \psi(x)\, dx$ to mean zero, whatever the nature of the function ψ may be.

By replacing ϕ by the upper limit of its absolute value, we see that $\Psi(x)$ remains finite, and hence that Φ approaches zero as x approaches a. Consequently Φ is continuous at a. On the other hand the same substitution shows that the integral Ψ converges uniformly when $a < x \leqq b$. For any fixed positive $\delta < 1$ the function

$$\Psi_1(x) = \int_0^{1-\delta} \frac{\phi[a + s(x-a)]}{(1-s)^\lambda}\, ds$$

is continuous throughout the interval $a < x \leqq b$, since the integrand in Ψ_1 is finite in the rectangle

$$a < x \leqq b, \quad 0 \leqq s \leqq 1 - \delta,$$

and is discontinuous only along a finite number of curves in this rectangle each of which is met by a line $x = x_0$ in at most one point. Since, as we have just seen, $\Psi_1(x)$ approaches $\Psi(x)$ uniformly as δ approaches zero, it follows from a fundamental theorem in uniform convergence that $\Psi(x)$ is continuous when $a < x \leqq b$, and hence the same is true of $\Phi(x)$, and our theorem is proved.

THEOREM 3. *If, in I, $\phi(x)$ is continuous and has a derivative which is finite and which has at most a finite number of discontinuities in I, and if $\phi(a) = 0$, the function*

$$\Phi(x) = \int_a^x \frac{\phi(\xi)}{(x-\xi)^\lambda}\, d\xi \qquad\qquad (\lambda < 1)$$

has a derivative continuous throughout I and given by the formula

$$\Phi'(x) = \int_a^x \frac{\phi'(\xi)}{(x-\xi)^\lambda}\, d\xi.$$

For if we integrate the expression for $\Phi(x)$ by parts, we have, when we remember that $\phi(a) = 0$,

$$\Phi(x) = \frac{1}{1-\lambda} \int_a^x (x-\xi)^{1-\lambda}\, \phi'(\xi)\, d\xi.$$

Applying here the rule for differentiating an integral whose limits are variable, we get the desired expression for $\Phi'(x)$. Hence from Theorem 2 it is evident that $\Phi'(x)$ is continuous. It should be noticed that when $\lambda > 0$ the integrals with which we have to deal are *infinite integrals* (i.e. integrals in which the integrand does not remain finite) so that the application to them of the ordinary rules of the calculus requires careful justification.

An alternative form of proof for this theorem consists in applying Dirichlet's Extended Formula* as follows:

$$\Phi(x) = \int_a^x \frac{1}{(x-\xi)^\lambda} \int_a^\xi \phi'(s)\,ds\,d\xi = \int_a^x \phi'(s) \int_s^x \frac{d\xi}{(x-\xi)^\lambda}\,ds$$

$$= \int_a^x \phi'(s) \int_s^x \frac{d\xi}{(\xi-s)^\lambda}\,ds = \int_a^x \int_a^\xi \frac{\phi'(s)}{(\xi-s)^\lambda}\,ds\,d\xi.$$

The differentiation of this last formula gives us the result we wish to establish †.

In conclusion we point out by means of the following two examples that if we replace the condition of finiteness for ϕ' by the condition of integrability, or even of absolute integrability, Φ will not always have a continuous derivative:

(1) $\phi(x) = (x-a)^\lambda$, $\qquad\qquad \Phi(x) = k(x-a)$,

(2) $\phi(x) = \begin{cases} 0 & (a \leqq x \leqq a') \\ (x-a')^\lambda & (a' < x \leqq b) \end{cases}$, $\quad \Phi(x) = \begin{cases} 0 & (a \leqq x \leqq a') \\ k(x-a')(a' < x \leqq b) \end{cases}$.

In both cases k is a positive constant, and if $0 < \lambda < 1$, ϕ is continuous in I and has a derivative which is continuous except at one point and absolutely integrable but not finite. In the first case Φ' is continuous, in the second discontinuous.

2. Abel's Mechanical Problem.
In one of his earliest published papers ‡ Abel showed how a certain mechanical problem, which includes the problem of the tautochrone as a special case, leads to what has since come to be called an integral equation, on whose solution the solution of the problem depends. On account of its great historical interest, we take up this problem in this section.

A particle starting from rest at a point P on a smooth curve which lies in a vertical plane, slides down the curve to its lowest point O. The velocity

* It should be noticed that we use this formula here under slightly different restrictions on the function $\phi(x, y)$ since ϕ is now a function of y alone, and therefore if it is discontinuous at all, is discontinuous along lines parallel to the axis of x.

† This method of reasoning admits of immediate extension to the proof of the more general formula

$$\frac{d}{dx} \int_a^x \psi(x-\xi)\phi(\xi)\,d\xi = \int_a^x \psi(x-\xi)\phi'(\xi)\,d\xi,$$

which holds under suitable restrictions on ψ.

‡ See *Collected Works*, p. 11. This paper was first published in Christiania in 1823. Cf. also a second paper beginning on p. 97 of the *Collected Works*, and originally published in *Crelle*, vol. 1 (1826), p. 153.

acquired at O will be independent of the shape of the curve. The time of descent T will however depend on this shape. We take O as origin, the axis of x vertically upward, and the axis of y horizontal and in the plane of the curve. Let the coordinates of the point of departure P be (x, y), and the coordinates of the point Q reached by the particle at the time t be (ξ, η), g the gravitational constant, and s the arc OQ. The velocity of the particle at Q is

$$-\frac{ds}{dt} = \sqrt{2g\,(x - \xi)}.$$

Hence
$$\sqrt{2g}\,t = -\int_P^Q \frac{ds}{\sqrt{x - \xi}}.$$

If we express s in terms of ξ

$$s = v\,(\xi),$$

the whole time of descent is then

$$T = \frac{1}{\sqrt{2g}} \int_0^x \frac{v'\,(\xi)\,d\xi}{\sqrt{x - \xi}}.$$

If the shape of the curve is given, the function v may be computed, and the whole time of descent is given to us as a function of x by the last formula.

The problem which Abel set himself is the converse of this, namely to determine the curve for which the time of descent is a given function of x. If we write

$$\sqrt{2g}\,T = f\,(x),$$

our problem is to determine the function v from the equation

$$f\,(x) = \int_0^x \frac{v'\,(\xi)\,d\xi}{\sqrt{x - \xi}} \qquad (1).$$

The formula for the solution of this integral equation was obtained by Abel by two different methods. The first depends on the use of series proceeding according to powers, not necessarily integral, of the argument; while the second, of a more general character, is closely related to the one we are about to give in the next section. Neither of Abel's methods can be regarded as satisfactory although they lead to the correct result. Among other objections it may be said that both methods omit the essential step of proving that the equation (1) has a solution.

3. Solution of Abel's Equation*. Instead of the equation (1) of § 2, Abel set himself the problem of solving a more general equation which we will write in the form

$$f(x) = \int_a^x \frac{u(\xi)\, d\xi}{(x-\xi)^\lambda} \qquad (0 < \lambda < 1) \qquad (1),$$

where f is a known function, u the function to be determined.

In order to solve (1) we begin by establishing the general formula (2) below. We start from the well-known formula

$$\frac{\pi}{\sin \mu\pi} = \int_\xi^z \frac{dx}{(z-x)^{1-\mu}(x-\xi)^\mu} \qquad (0 < \mu < 1).$$

Let $\phi(\xi)$ be any function which is continuous and has a continuous derivative throughout I. Multiply this equation by $\phi'(\xi)\, d\xi$ and integrate from a to z, which we suppose to be any point of I. This gives

$$\frac{\pi}{\sin \mu\pi}[\phi(z) - \phi(a)] = \int_a^z \int_\xi^z \frac{\phi'(\xi)}{(z-x)^{1-\mu}(x-\xi)^\mu}\, dx\, d\xi.$$

If we apply Dirichlet's Generalized Formula to the second member of this equation, we get the desired result

$$\phi(z) - \phi(a) = \frac{\sin \mu\pi}{\pi} \int_a^z \frac{1}{(z-x)^{1-\mu}} \int_a^x \frac{\phi'(\xi)\, d\xi}{(x-\xi)^\mu}\, dx \qquad (2),$$

a formula which holds under the sole restrictions that z be in I, and ϕ be continuous and have a continuous derivative in I, and that

$$0 < \mu < 1.$$

Theorem 2, § 1, shows us at once that a necessary condition that (1) have a solution continuous throughout I is that $f(x)$ be continuous throughout I and that $f(a) = 0$.

Let us suppose that these conditions are fulfilled and that $u(x)$ is a continuous solution of (1). Multiply (1) by $(z-x)^{\lambda-1}\, dx$, where z is a point of I, and integrate from a to z, thus getting

$$\int_a^z \frac{f(x)\, dx}{(z-x)^{1-\lambda}} = \int_a^z \frac{1}{(z-x)^{1-\lambda}} \int_a^x \frac{u(\xi)\, d\xi}{(x-\xi)^\lambda}\, dx.$$

If in (2) we let

$$\phi(x) = \int_a^x u(\xi)\, d\xi,$$

---

* Except for the method of deducing formula (2), the method we use is, barring notation, that of Liouville in *Liouville's Journal*, vol. 4 (1839), p. 233. Liouville, who seems not to have been aware of Abel's work, had already published on this subject in the *Journal de l'École Polytechnique*, Cahier 21 (1832), p. 1.

it will be seen that the preceding equation reduces to

$$\int_a^z \frac{f(x)\,dx}{(z-x)^{1-\lambda}} = \frac{\pi}{\sin \lambda\pi} \int_a^z u(\xi)\,d\xi \qquad (3).$$

Since the second member of (3) has a continuous derivative with regard to z, the same must be true of the first member, and this gives us a further necessary condition for (1) having a continuous solution. By differentiating (3), we get as the value of this solution

$$u(z) = \frac{\sin \lambda\pi}{\pi} \frac{d}{dz} \int_a^z \frac{f(x)\,dx}{(z-x)^{1-\lambda}} \qquad (4).$$

We thus see that u is completely determined, that is that (1) does not have more than one continuous solution. That the formula (4) really does give a solution of (1) may be seen by substituting it in (1). The second member of (1) thus becomes

$$\frac{\sin \lambda\pi}{\pi} \int_a^x \frac{1}{(x-\xi)^\lambda} \frac{d}{d\xi} \int_a^\xi \frac{f(x)\,dx}{(\xi-x)^{1-\lambda}} \, d\xi,$$

which reduces by means of Theorem 3, § 1, to

$$\frac{\sin \lambda\pi}{\pi} \frac{d}{dx} \int_a^x \frac{1}{(x-\xi)^\lambda} \int_a^\xi \frac{f(x)\,dx}{(\xi-x)^{1-\lambda}} \, d\xi,$$

and this in turn reduces by means of (2), when we let

$$\phi(z) = \int_a^z f(x)\,dx,$$

to
$$\frac{d}{dx} \int_a^x f(z)\,dz = f(x).$$

Thus we see that (4) is a solution of (1), and we have proved

THEOREM 1. *A necessary and sufficient condition that (1) have a solution continuous in I is that $f(x)$ be continuous in I, that $f(a)=0$, and that*

$$\int_a^x \frac{f(\xi)\,d\xi}{(x-\xi)^{1-\lambda}}$$

have a continuous derivative throughout I. If these conditions are fulfilled, (1) has only one continuous solution, given by formula (4).

An important case in which these conditions are fulfilled is that in which f is continuous and has a derivative which is finite, and has at most a finite number of discontinuities in I, and $f(a)=0$. This we see from Theorem 3, § 1, from which we also see that in this case (4) may be written

$$u(z) = \frac{\sin \lambda\pi}{\pi} \int_a^z \frac{f'(x)\,dx}{(z-x)^{1-\lambda}} \qquad (5).$$

Hence

THEOREM 2. *If $f(x)$ is continuous and has a derivative finite in I and with only a finite number of discontinuities there, and $f(a) = 0$, equation* (1) *has one and only one continuous solution, and this is given by formula* (5) *.

While this is essentially Abel's result, that mathematician did not consider the integral equation (1) but rather the differentio-integral equation

$$f(x) = \int_a^x \frac{v'(\xi)\, d\xi}{(x-\xi)^\lambda} \qquad (0 < \lambda < 1) \qquad (6).$$

By means of the theorems just established and Theorems 2, 3 of § 1 we readily deduce the result

THEOREM 3. *A necessary and sufficient condition that* (6) *have a solution which together with its derivative is continuous throughout I is that $f(x)$ be continuous in I, that $f(a) = 0$, and that*

$$\int_a^x \frac{f(\xi)\, d\xi}{(x-\xi)^{1-\lambda}}$$

have a continuous derivative throughout I. If these conditions are fulfilled, the general solution of (6) *is*

$$v(z) = k + \frac{\sin \lambda \pi}{\pi} \int_a^z \frac{f(x)\, dx}{(z-x)^{1-\lambda}},$$

where k is an arbitrary constant.

By letting $\lambda = \frac{1}{2}$ we get the solution of the mechanical problem of § 2. If in particular we let $f(x) = \text{const.}$, we get Abel's solution of the problem of the tautochrone.

An easy extension of the results we have found is to the case in which

---

* Goursat, in *Acta Math.* vol. 27 (1903), pp. 131—133, has shown that equation (1) still has a solution, though not a continuous one, if we drop the requirement that $f(a) = 0$. This may be readily seen by bringing in, in place of u, the function

$$v(x) = u(x) - \frac{\sin \lambda \pi}{\pi} \frac{f(a)}{(x-a)^{1-\lambda}}.$$

Making this substitution, we find that equation (1) reduces to

$$f(x) - f(a) = \int_a^x \frac{v(\xi)\, d\xi}{(x-\xi)^\lambda}.$$

Conversely, we see that a solution of this last equation corresponds to a solution of (1). Consequently a solution of (1) is

$$u(z) = \frac{\sin \lambda \pi}{\pi} \frac{f(a)}{(z-a)^{1-\lambda}} + \frac{\sin \lambda \pi}{\pi} \int_a^z \frac{f'(x)\, dx}{(z-x)^{1-\lambda}}.$$

λ is negative, cf. Liouville, *loc. cit.* Let us for instance suppose that $-1 < \lambda < 0$. We see by differentiation that any continuous solution of (1) also satisfies the equation

$$f'(x) = -\lambda \int_a^x \frac{u(\xi)\,d\xi}{(x-\xi)^{\lambda+1}} \qquad (7),$$

and conversely, if $f(x)$ is continuous in I and $f(a) = 0$, we see by integrating that every continuous solution of (7) is also a solution of (1). Hence

THEOREM 4. *If $0 > \lambda > -1$, a necessary and sufficient condition that* (1) *have a continuous solution is that $f(x)$ be continuous together with its first derivative throughout I, that $f(a) = f'(a) = 0$, and that*

$$\int_a^x \frac{f'(\xi)\,d\xi}{(x-\xi)^{-\lambda}}$$

have a continuous derivative throughout I. If these conditions are fulfilled, (1) *has one and only one continuous solution, namely*

$$u(x) = \frac{\sin \lambda\pi}{\pi} \frac{d}{dz} \int_a^z \frac{f'(x)\,dx}{(z-x)^{-\lambda}}.$$

In particular, the above conditions are fulfilled if $f(x)$, $f'(x)$, $f''(x)$ are continuous in I and $f(a) = f'(a) = 0$. In this case the continuous solution may be written

$$u(x) = \frac{\sin \lambda\pi}{\pi} \int_a^z \frac{f''(x)\,dx}{(z-x)^{-\lambda}}.$$

The extension to other cases in which λ is negative may readily be made by further differentiation. We leave to the reader the enunciation and proof of the general theorem to be obtained here as well as the consideration of the special cases in which λ is an integer negative or zero.

4. Liouville's Introduction of Integral Equations of the Second Kind. After Abel, the next to use an integral equation was Liouville, who, in the year 1837[*], showed that a particular solution of a certain linear differential equation can be obtained by solving an integral equation of a type somewhat different from, though closely resembling, Abel's equation.

To explain this we begin with the non-homogeneous equation

$$\frac{d^2y}{dx^2} + \rho^2 y = \phi(x) \qquad (1),$$

[*] *Liouville's Journal*, vol. 2, p. 24. This reference has no connection with Liouville's work cited in the preceding section.

where ρ is a parameter, and $\phi\,(x)$ a function continuous throughout I. The general solution of the reduced equation

$$\frac{d^2y}{dx^2} + \rho^2 y = 0$$

is $a \sin \rho\,(x-a) + \beta \cos \rho\,(x-a)$, so that by a well-known formula* the general solution of (1) is

$$y\,(x) = a \sin \rho\,(x-a) + \beta \cos \rho\,(x-a) + \frac{1}{\rho} \int_a^x \phi\,(\xi) \sin \rho\,(x-\xi)\,d\xi \quad (2).$$

Now consider the homogeneous equation

$$\frac{d^2y}{dx^2} + [\rho^2 - \sigma\,(x)]\,y = 0 \qquad (3),$$

where $\sigma\,(x)$ is continuous in I. We will denote by $u\,(x)$ the solution of (3) which satisfies the auxiliary conditions

$$u\,(a) = 1, \quad u'\,(a) = 0 \qquad (4).$$

This function will also be a solution of the non-homogeneous equation

$$\frac{d^2y}{dx^2} + \rho^2 y = \sigma\,(x)\,u\,(x).$$

Consequently by using (2) and (4) we get

$$u\,(x) = \cos \rho\,(x-a) + \frac{1}{\rho} \int_a^x \sigma\,(\xi) \sin \rho\,(x-\xi)\,u\,(\xi)\,d\xi \qquad (5).$$

Conversely if $u\,(x)$ is any continuous solution of this integral equation, we see, by applying to (5) the rule for differentiating an integral whose limits are functions of x, that $u\,(x)$ has continuous first and second derivatives given by the formulae

$$u'\,(x) = -\rho \sin \rho\,(x-a) + \int_a^x \sigma\,(\xi) \cos \rho\,(x-\xi)\,u\,(\xi)\,d\xi \qquad (6),$$

$$u''\,(x) = -\rho^2 \cos \rho\,(x-a) - \rho \int_a^x \sigma\,(\xi) \sin \rho\,(x-\xi)\,u\,(\xi)\,d\xi + \sigma\,(x)\,u\,(x) \quad (7).$$

Formulae (5) and (7) show that u satisfies (3), while from (5) and (6) we see that conditions (4) are satisfied. Thus, since there is only one function satisfying (3) and (4), we have proved the

THEOREM. *The solution of the differential equation* (3) *which satisfies the auxiliary conditions* (4) *is a solution of the integral equation* (5); *and conversely this integral equation* (5) *has only one continuous solution.*

The possibility here illustrated of replacing a differential equation

* Cf. Forsyth's *Treatise on Differential Equations*, § 66.

together with certain auxiliary conditions by a single integral equation is characteristic for many important applications of integral equations.

Liouville's object in introducing the integral equation (5) was to enable him to obtain a development of $u\,(x)$ in a series which converges rapidly for large values of ρ. Such a series he obtained by solving (5) by a method which will be explained in the next section.

Comparing Abel's equation (formula (1), § 3) with Liouville's equation (5), we see that they come respectively under the following types:

$$f\,(x) = \int_{a}^{x} K\,(x,\,\xi)\,u\,(\xi)\,d\xi \qquad\qquad (8),$$

$$u\,(x) = f\,(x) + \int_{a}^{x} K\,(x,\,\xi)\,u\,(\xi)\,d\xi \qquad\qquad (9),$$

in which $f\,(x)$ and $K\,(x,\,\xi)$ are to be regarded as known functions and $u\,(x)$ is the function to be determined. Equations (8) and (9) are spoken of as linear integral equations of the first and second kinds respectively. K is called the *kernel* of these equations*.

In place of the equations (8) and (9) in which the upper limit of the integrals is the variable x we often have to deal with equations of exactly the same form in which the upper limit is the constant b. These are also called linear integral equations of the first and second kind respectively. It will be seen that equations (8) and (9) are merely the special cases of the equations just mentioned in which the kernel $K\,(x,\,\xi)$ vanishes when $\xi > x$ since it then makes no difference whether x or b be used as the upper limit of integration.

The special case of (9), or of the more general equation in which the upper limit of integration is b, in which $f\,(x)$ vanishes identically may be called a *homogeneous* integral equation of the second kind. It should not be confounded with the equation of the first kind.

5. The Method of Successive Substitutions†. The method which Liouville used for solving equation (5) of the last section applies

---

* These terms were first employed by Hilbert, *Göttinger Nachrichten*, 1904. An equation of the form

$$\psi\,(x)\,u\,(x) = f\,(x) + \int_{a}^{b} K\,(x,\,\xi)\,u\,(\xi)\,d\xi,$$

which includes the equations of the first and second kinds as special cases, has been called by Hilbert an equation of the third kind, but only a very special case of this equation has so far been treated.

† This method of solving an integral equation of the second kind is usually attributed to C. Neumann, whose work, however, is more than thirty years later than Liouville's. The connection of Liouville's work with the theory of integral equations of the second kind has been generally overlooked. For a formulation of the method of successive substitutions which is convenient even in far more complicated cases, cf. Mason, *Math. Ann.* vol. 65 (1908), p. 570.

with equal simplicity to the more general equation (9). In fact we will consider at once the still more general equation referred to near the end of the last section

$$u(x) = f(x) + \int_a^b K(x, \xi) u(\xi) d\xi \qquad (1).$$

We assume that the kernel K is finite in S and that its discontinuities, if it has any, are regularly distributed there*.

If (1) is to have a solution continuous in I, it is clear (cf. Theorem 1, § 1) that $f(x)$ must be continuous in I. Let us assume that this condition is fulfilled.

Assuming that (1) has a continuous solution, we now proceed to find it. Substitute in the second member for $u(\xi)$ the value given by the equation itself, thus getting

$$u(x) = f(x) + \int_a^b K(x, \xi) f(\xi) d\xi + \int_a^b K(x, \xi) \int_a^b K(\xi, \xi_1) u(\xi_1) d\xi_1 d\xi.$$

Here we again substitute for $u(\xi_1)$ its value as given by (1), and thus get a four-term expression for $u(x)$. Proceeding in this way, we get the general formula

$$u(x) = S_n(x) + R_n(x) \qquad (2),$$

where

$$S_n(x) = f(x) + \int_a^b K(x, \xi) f(\xi) d\xi + \int_a^b K(x, \xi) \int_a^b K(\xi, \xi_1) f(\xi_1) d\xi_1 d\xi$$

$$+ \dots + \int_a^b K(x, \xi) \int_a^b K(\xi, \xi_1) \dots \int_a^b K(\xi_{n-2}, \xi_{n-1}) f(\xi_{n-1}) d\xi_{n-1} \dots d\xi_1 d\xi,$$

$$R_n(x) = \int_a^b K(x, \xi) \int_a^b K(\xi, \xi_1) \dots \int_a^b K(\xi_{n-1}, \xi_n) u(\xi_n) d\xi_n \dots d\xi_1 d\xi.$$

We may regard this expression for $u(x)$ as a *finite* series of $n + 1$ terms plus a remainder, this remainder, however, involving the very function $u(x)$ which we are developing†. This suggests to us the consideration of the infinite series

$$f(x) + \int_a^b K(x, \xi) f(\xi) d\xi + \int_a^b K(x, \xi) \int_a^b K(\xi, \xi_1) f(\xi_1) d\xi_1 d\xi + \dots \ (3).$$

From Theorem 1, § 1, we see that the terms of this series are continuous in I. The series therefore represents a function continuous throughout I provided it converges uniformly there. We will prove that whenever this is the case, the function $u(x)$ represented by (3) is a solution of (1). For this purpose multiply the series for $u(\xi)$ by

---

* In the work of this section and the next it is immaterial whether the functions f and K are assumed to be real or are allowed to be complex.

† Cf. the various forms of Taylor's series with a remainder.

$K(x, \xi)$ and integrate the resulting series with regard to ξ from a to b term by term, as we have a right to do on account of its uniform convergence. This gives us precisely series (3) without the first term; that is

$$\int_a^b K(x, \xi) u(\xi) d\xi = u(x) - f(x),$$

and this is the integral equation of which we wished to prove that $u(x)$ is a solution.

We will now obtain two different sufficient conditions for the uniform convergence of (3).

Let us first consider the case in which

$$K(x, \xi) = 0 \text{ when } \xi > x \qquad (4).$$

This, as we have seen, is the case in which the upper limit of integration in (1) may be taken as x. Here the general term of the series (3) may be written

$$F_n(x) = \int_a^x K(x, \xi) \int_a^\xi K(\xi, \xi_1) \dots \int_a^{\xi_{n-1}} K(\xi_{n-1}, \xi_n) f(\xi_n) d\xi_n \dots d\xi_1 d\xi.$$

Let M and N be constants such that

$$|K(x, \xi)| < M, \quad |f(x)| < N \qquad (a \leq \xi \leq x \leq b).$$

Then

$$|F_n(x)| \leq NM^{n+1} \int_a^x \int_a^\xi \dots \int_a^{\xi_{n-1}} d\xi_n \dots d\xi_1 d\xi = NM^{n+1} \frac{(x-a)^{n+1}}{(n+1)!}$$

$$\leq N \frac{[M(b-a)]^{n+1}}{(n+1)!} \qquad (a \leq x \leq b).$$

The series of which this last written positive constant is the general term is convergent. Consequently the series (3) is absolutely and uniformly convergent throughout I.

Without the special restriction we just made on K, the inequality which we should get in the same way for the general term $F_n(x)$ of (3) would be

$$|F_n(x)| \leq NM^{n+1} \int_a^b \int_a^b \dots \int_a^b d\xi_n \dots d\xi_1 d\xi = NM^{n+1} (b-a)^{n+1},$$

and the series of which this is the general term converges only when

$$M(b-a) < 1 \qquad (5).$$

Thus we see that (3) converges absolutely and uniformly when either one of the two conditions (4) or (5) is fulfilled.

We will now prove that in either of these cases the equation (1) cannot have more than one continuous solution. For this purpose we turn to formula (2) by which any continuous solution of (1) is expressed.

Referring to the formula for $R_n(x)$, we see that, if we denote by N' the maximum of $|u(x)|$ in I, the same reasoning which led us before to the inequalities for $|F_n(x)|$ now gives

$$|R_n(x)| \leqq N'\frac{[M(b-a)]^{n+1}}{(n+1)!} \text{ if } K(x,\xi)=0 \text{ when } \xi > x,$$
$$|R_n(x)| \leqq N'[M(b-a)]^{n+1} \text{ if } M(b-a) < 1.$$

Thus we see in either case that

$$\lim_{n=\infty} R_n(x) = 0.$$

On the other hand, $S_n(x)$ is simply the sum of the first $n+1$ terms of (3). Consequently the function u given by (2) is precisely the value of (3), and we have the two theorems :

THEOREM 1. *If $K(x, \xi)$ is finite in T and its discontinuities, if it has any, are regularly distributed, a necessary and sufficient condition that the equation*

$$u(x) = f(x) + \int_a^x K(x, \xi)\, u(\xi)\, d\xi \qquad (6)$$

have a solution continuous throughout I is that $f(x)$ be continuous throughout I, and if this condition is fulfilled, (6) has only one continuous solution, which is given by the absolutely and uniformly convergent series (3).

THEOREM 2. *If $K(x, \xi)$ is finite in S and its discontinuities, if it has any, are regularly distributed, and $f(x)$ is continuous in I, then provided that*

$$M(b-a) < 1,$$

where M denotes the upper limit of $|K|$ in S, equation (1) has one and only one solution continuous in I, and this solution is given by the absolutely and uniformly convergent series (3).

Besides the continuous solution whose existence has just been established, the integral equation of the second kind may also have discontinuous solutions. In order to show this, we will consider the special case

$$a = 0, \quad f(x) = 0, \quad K(x, \xi) = \xi^{x-\xi}.$$

The integral equation (6) then reduces to

$$u(x) = \int_0^x \xi^{x-\xi}\, u(\xi)\, d\xi \qquad (7).$$

If we take b as any positive constant, the conditions of Theorem 1 are fulfilled. Equation (7) has therefore one and only one continuous solution, and this solution is readily seen to be $u = 0$; as, indeed, is the case for every homogeneous integral equation of the second kind. By direct substitution, we verify that it also has the infinite number of discontinuous solutions

$$u(x) = kx^{x-1},$$

where k is an arbitrary constant different from zero.

It should be noticed that these solutions become not merely infinite when $x = 0$, but become infinite so strongly that $\int_0^b u(x)\,dx$ diverges. It may readily be seen that if the integral equation satisfies all the conditions of Theorem 1 or 2, any discontinuous solution which it may have will necessarily be non-integrable*.

These non-integrable solutions have not as yet proved to be of any importance and we shall not be concerned with them in this tract.

There are various extensions of the considerations of this section which naturally suggest themselves. In the first place we may ask under what conditions it is possible to assert that the continuous solution of (1) or (6) has a continuous derivative. Let us assume that both $K(x, \xi)$ and $\partial K/\partial x$ are finite in S and that any discontinuities which these functions may have are regularly distributed. Also that the discontinuities of K lie on curves without double points, which have continuously turning tangents nowhere parallel to the axis of x or of ξ. It may readily be inferred from this that, on the curves where it is discontinuous, except perhaps at the points where these curves meet the boundary of S, the function K joins on continuously to continuous boundary values,—different values, of course, on the two sides of the curve. We may then, proceeding with some caution, differentiate equation (1) and thus arrive at the conclusion that under the conditions just stated a necessary and sufficient condition that the continuous solution u of (1) have a derivative continuous in the open interval $a < x < b$, is that $f(x)$ have a derivative continuous in this same interval. It would of course be possible to arrive at a similar result with much less drastic restrictions on K.

A second question which may be raised is as to the existence of continuous solutions of (1) or (6) when the kernel is not finite. We will consider here only a special case which may serve as a sample of others.

---

* We note in passing that an integral equation of the *first* kind may, under similar conditions, have a discontinuous solution which is integrable. Cf. the footnote to Theorem 2, § 3.

In the equation (6), let K be defined throughout T so that

$$K(x, \xi) = \frac{G(x, \xi)}{(x - \xi)^{\lambda}} \qquad (0 < \lambda < 1),$$

where G is finite in T and where its discontinuities, if any exist, are regularly distributed. It may now be readily proved that if $\psi(x)$ is any function continuous in I, the function

$$\int_a^x K(x, \xi)\, \psi(\xi)\, d\xi$$

is also continuous throughout I.

If then $f(x)$ is continuous in I, we see that the same is true of each term of the series (3), in which it must be remembered that the upper limits of integration are now variable.

For the general term $F_n(x)$ of this series we readily get the inequality

$$|F_n(x)| \leqq NM^{n+1} \int_a^x \frac{1}{(x - \xi)^{\lambda}} \int_a^{\xi} \frac{1}{(\xi - \xi_1)^{\lambda}} \cdots \int_a^{\xi_{n-1}} \frac{1}{(\xi_{n-1} - \xi_n)^{\lambda}}\, d\xi_n \cdots d\xi,$$

where
$$|f(x)| \leqq N, \qquad |G(x, \xi)| \leqq M.$$

In terms of the quantities

$$k_p = \int_0^1 \frac{s^{p(1-\lambda)}}{(1 - s)^{\lambda}}\, ds$$

the $(n + 1)$-fold integral last written may be readily evaluated, and we thus find

$$|F_n(x)| \leqq NM^{n+1}k_0 k_1 \cdots k_n (x - a)^{(n+1)(1-\lambda)}.$$

If, then, we can prove that the series whose general term is

$$NM^{n+1}k_0 k_1 \cdots k_n (b - a)^{(n+1)(1-\lambda)} \qquad (8)$$

converges, the absolute and uniform convergence of (3) follows. The ratio of two successive terms of the form (8) is

$$M(b - a)^{1-\lambda}k_n.$$

The convergence therefore follows if we can show that

$$\lim_{n = \infty} k_n = 0,$$

and this property of the k's may be established with ease from their definition.

Having thus established the uniform convergence of (3), the proof that this series represents a continuous solution of the integral equation, and the further proof that this is the only continuous solution of this equation follow almost as before, and we thus get

THEOREM 3.   *The equation*

$$u(x) = f(x) + \int_a^x \frac{G(x, \xi)}{(x - \xi)^\lambda} u(\xi) d\xi \qquad (0 < \lambda < 1)$$

in which f is continuous in I, and G is finite in T and such discontinuities as it may have are regularly distributed, has one and only one continuous solution, and this solution is given by the method of successive substitutions in the form of an absolutely and uniformly convergent series.

6. Volterra's Treatment of Equations of the Second Kind. Iterated and Reciprocal Functions.

We will now start afresh and approach the theory of integral equations of the second kind from a new point of view due to Volterra*.

Let us assume that $K(x, y)$ is finite in S and that any discontinuities which it may have are regularly distributed. From K we form the *iterated functions* $K_1, K_2, \ldots$ by means of the formulae

$$\left.\begin{aligned} K_1(x, y) &= K(x, y) \\ K_i(x, y) &= \int_a^b K(x, \xi) K_{i-1}(\xi, y) d\xi \end{aligned}\right\} \qquad (1).$$

A reference to Theorem 1, § 1, shows us that all of these functions, except possibly K_1, are continuous throughout S.

By successive applications of (1) we get

$$K_i(x, y) = \int_a^b \cdots \int_a^b K(x, \xi_1) K(\xi_1, \xi_2) \cdots K(\xi_{i-1}, y) d\xi_{i-1} \cdots d\xi_1 \qquad (2).$$

If, by means of (2), we express $K_i(x, \xi)$ and $K_j(\xi, y)$ as $(i - 1)$ and $(j - 1)$-fold integrals respectively, we get for $\int_a^b K_i(x, \xi) K_j(\xi, y) d\xi$ an $(i + j - 1)$-fold integral, which, when we change the order of integration, becomes, except for notation, precisely the value of $K_{i+j}(x, y)$ given by (2). We have thus established the important formula

$$K_{i+j}(x, y) = \int_a^b K_i(x, \xi) K_j(\xi, y) d\xi \qquad (i, j = 1, 2, \ldots) \qquad (3)$$

which includes definition (1) as a special case.

---

* Six papers come primarily into consideration here and in the following sections,—four in the *Atti* of the Turin Academy, vol. 31, 1896 (Jan. 12—April 26), and two in the *Rendiconti* of the *Accademia dei Lincei*, series 5, vol. 5^1, 1896 (March, April). These articles will be referred to as *Torino* I, II, III, IV and *Lincei* I, II, respectively. There is also a paper by the same author in the *Annali di Matematica*, ser. 2, vol. 25 (1897), p. 139, which should be cited here. It may be mentioned that Volterra's work on integral equations commenced in 1884 with a note "Sopra un Problema di Elettrostatica," *Atti d. R. Accad. dei Lincei*, ser. 3, Transunti, vol. 8, p. 315, in which a relation to the calculus of variations was pointed out. For the present section, cf. *Lincei* I.

If $K(x, y)$ vanishes when $y > x$, we see from (1) that $K_i(x, y)$ also vanishes when $y > x$ for all values of i. Consequently in this case (3) may be written

$$K_{i+j}(x, y) = \int_y^x K_i(x, \xi) K_j(\xi, y) \, d\xi \qquad (i, j = 1, 2, \ldots) \quad (4).$$

The case just mentioned is the only one considered by Volterra, and the formula used by him is formula (4).

Returning to the general case, let us consider the series

$$K_1(x, y) + K_2(x, y) + K_3(x, y) + \ldots \tag{5}.$$

It will be shown presently that in certain important cases this series converges uniformly throughout S. Let us assume that this is the case, and denote the value of (5) by $-k(x, y)$. Since every term in (5), except possibly the first, is continuous in S, it follows from the assumed uniform convergence of (5) that $k(x, y)$ is finite in S and is discontinuous only where $K(x, y)$ is discontinuous.

Let us denote the sum of the first n terms of (5) by $S_n(x, y)$ and the remainder after this point by $R_n(x, y)$, so that

$$R_n(x, y) = \sum_{i=n+1}^{\infty} \int_a^b K_{i-j}(x, \xi) K_j(\xi, y) \, d\xi \tag{6}.$$

In this formula, the integer j may, if we please, vary from term to term, being however always less than i.

Let us first assign to j the value n in all the terms. Then the series in (6) is what we should get by starting from the series

$$-k(x, \xi) = \sum_{i=n+1}^{\infty} K_{i-n}(x, \xi),$$

multiplying it by $K_n(\xi, y)$ and integrating it term by term. Since this integration is allowable, we get

$$R_n(x, y) = -\int_a^b k(x, \xi) K_n(\xi, y) \, d\xi \tag{7}.$$

On the other hand, let us give to j in (6) the value $i - n$. Then the series in (6) may be obtained by starting from the series

$$-k(\xi, y) = \sum_{i=n+1}^{\infty} K_{i-n}(\xi, y),$$

multiplying it by $K_n(x, \xi)$ and integrating it term by term. We thus get as a second form for the remainder

$$R_n(x, y) = -\int_a^b \dot{K}_n(x, \xi) k(\xi, y) \, d\xi \tag{8}.$$

Since we have

$$R_1(x, y) = -k(x, y) - K(x, y),$$

EQUATIONS OF THE SECOND KIND

it follows that in the case $n = 1$ the two formulae (7), (8) may be combined into

$$K(x, y) + k(x, y) = \int_a^b K(x, \xi) k(\xi, y) \, d\xi$$
$$= \int_a^b k(x, \xi) K(\xi, y) \, d\xi \qquad (9).$$

This is one of the most fundamental formulae in the theory of integral equations. We base upon it the following

DEFINITION. *Two functions $K(x, y)$ and $k(x, y)$ are said to be reciprocal if they are finite in S, if any discontinuities they may have are regularly distributed, and if they satisfy* (9)*.

The relation here defined between two functions is, as the very name "reciprocal" implies, independent of the order in which these functions are taken, since (9) is not changed by an interchange of K and k. On the other hand, we see from (9) (cf. Theorem 1, § 1) that the sum of two reciprocal functions is continuous.

We are now in a position to give Volterra's elegant solution of the integral equation of the second kind

$$u(x) = f(x) + \int_a^b K(x, \xi) u(\xi) \, d\xi \qquad (10).$$

Here we assume as before that K is finite in S and that any discontinuities it may have are regularly distributed. We also assume that there exists a function $k(x, y)$ reciprocal to $K(x, y)$.

If (10) has a continuous solution $u(x)$, we may write

$$u(\xi) = f(\xi) + \int_a^b K(\xi, \xi_1) u(\xi_1) \, d\xi_1.$$

Multiplying this equation by $k(x, \xi)$ and integrating, we get

$$\int_a^b k(x, \xi) u(\xi) \, d\xi$$
$$= \int_a^b k(x, \xi) f(\xi) \, d\xi + \int_a^b \int_a^b k(x, \xi) K(\xi, \xi_1) u(\xi_1) \, d\xi_1 \, d\xi \qquad (11).$$

If we reverse the order of integration, and then apply (9), the second term on the right becomes

$$\int_a^b [K(x, \xi) + k(x, \xi)] u(\xi) \, d\xi.$$

---

* We note, for later use, the following application of this definition, which was given by Goursat (*C. R.* Feb. 17, 1908): If K and k are reciprocal, and if $r(x)$ is continuous in I and does not vanish there, then

$$\frac{r(x)}{r(y)} K(x, y) \text{ and } \frac{r(x)}{r(y)} k(x, y)$$

are also reciprocal.

The second part of this cancels against the first member of (11), while the first part may be evaluated by means of (10). Thus (11) reduces to

$$u\,(x) = f\,(x) - \int_a^b k\,(x,\,\xi)\,f\,(\xi)\,d\xi \qquad (12).$$

We see then that, under the conditions we have imposed, (10) cannot have more than one continuous solution, and if it has one, this will be given by formula (12).

In order to prove that the continuous function $u\,(x)$ defined by (12) really is a solution of (10), let us write (12) in the form

$$f\,(x) = u\,(x) + \int_a^b k\,(x,\,\xi)\,f\,(\xi)\,d\xi.$$

This may be regarded as an integral equation for determining $f\,(x)$. Since K is a function reciprocal to k, we see that this equation satisfies all the conditions which we previously imposed on (10), so that, by what we have just proved, its continuous solution $f\,(x)$ is given by the formula

$$f\,(x) = u\,(x) - \int_a^b K\,(x,\,\xi)\,u\,(\xi)\,d\xi.$$

This, however, is precisely the equation (10), which we thus see is satisfied by $u\,(x)$, and we have proved the

THEOREM 1. *If $K\,(x,\,\xi)$ is finite in S and any discontinuities it may have are regularly distributed, and $f\,(x)$ is continuous in I, then the equation (10) has one and only one continuous solution provided there exists a function $k\,(x,\,y)$ reciprocal to $K\,(x,\,y)$; and in this case this solution is given by (12).*

We saw above that a function reciprocal to K will exist provided the series (5) converges uniformly. Although this is, as we shall see in § 9 (Theorems 5, 6) by no means a necessary condition for the existence of a reciprocal function, it will nevertheless be of interest to determine certain cases in which this condition is fulfilled.

THEOREM 2*. *If $K\,(x,\,y)$ is finite in S and any discontinuities it may have are regularly distributed, there will exist a reciprocal function given by series (5) provided that*

$$M\,(b - a) < 1,$$

where M is the upper limit of $|\,K\,(x,\,y)\,|$ in S.

---

* For a theorem which in some cases is more far-reaching than this, see Theorem 5, § 12.

For under these conditions, as we see from (2),
$$| K_i (x, y) | \leqq M^i (b - a)^{i-1}$$
and from this the absolute and uniform convergence of (5) follows at once.

THEOREM 3. *If $K(x, y)$ is finite in S and any discontinuities it may have are regularly distributed, there will exist a reciprocal function given by series (5) provided that*
$$K(x, y) = 0 \qquad\qquad when\ y > x.$$

In this case we can use formula (4), by means of which we easily establish the inequality
$$| K_i (x, y) | \leqq \frac{M^i (x-y)^{i-1}}{(i-1)!} \qquad (y < x).$$

Consequently the general term of (5) does not exceed in absolute value the quantity
$$\frac{M^i (b - a)^{i-1}}{(i-1)!} ,$$
and since this is the general term of a convergent series of positive constant terms, the absolute and uniform convergence of (5) follows at once.

In either of these two cases, or indeed in any case in which the series (5) converges uniformly, the solution (12) may be written
$$u(x) = f(x) + \sum_{n=1}^{\infty} \int_a^b K_n (x, \xi) f(\xi)\, d\xi.$$

If here we replace the iterated function K_n by its value (2), we get by a mere change in the order of integration precisely the series which we found in § 5 by the method of successive substitutions. Thus we get a new proof of Theorems 1, 2, § 5. Conversely, from those theorems all the results of this section can readily be deduced so far as they relate to the cases covered by those theorems.

Finally we prove

THEOREM 4. *There cannot exist two different functions $k_1(x, y)$ and $k_2(x, y)$ both reciprocal to the same function $K(x, y)$.*

For, as we have seen, $K + k_1$ and $K + k_2$ would both be continuous; accordingly the same would be true of the difference of these two functions
$$\sigma(x, y) = k_1(x, y) - k_2(x, y).$$

By substituting in (9) first $k = k_1$, then $k = k_2$, and subtracting the resulting equations from each other, we find
$$\sigma(x, y) = \int_a^b K(x, \xi)\, \sigma(\xi, y)\, d\xi \qquad\qquad (13).$$

This may be regarded as a homogeneous integral equation of the second kind for determining σ, y being regarded as a parameter. Since, by hypothesis, K has a reciprocal function, all the conditions of Theorem 1 are fulfilled. Consequently (13) has only one continuous solution, and this is, by inspection, $\sigma = 0$. We thus have $k_1 = k_2$, and the assumption of two *different* functions reciprocal to K is impossible.

7. Linear Algebraic Equations with an Infinite Number of Variables. We come now to a very remarkable and important relation between the theory of integral and of algebraic equations. This relation seems to have been first noticed by Volterra, who pointed out* that an integral equation of the *first* kind may be regarded as the limiting form of a system of n linear algebraic equations in n variables as n becomes infinite. Though not explicitly mentioned by Volterra, his remarks make it at once clear that the same is true for equations of the *second* kind.

It is Fredholm's great achievement to have seen how this observation could be utilized to pass from the solution of the system of linear equations to the solution of the integral equation of the second kind, which he was thus enabled to treat in a far more general manner than it had ever been treated before†. This method was used by Fredholm as a heuristic method for discovering first the facts, and secondly methods by which they may be proved. Later Hilbert‡ showed how the theory can be rigorously established by following out in detail the limiting process of Volterra and Fredholm. We follow Fredholm, giving in this section merely the heuristic part of the work, and reserving the proofs for the next section. The whole of the present section is therefore, from a strictly logical point of view, superfluous and may be omitted.

Let us divide the interval ab into n equal parts by the points

$$x_1 = a + \delta, \quad x_2 = a + 2\delta, \quad \dots \quad x_n = a + n\delta = b \quad \left(\delta = \frac{b-a}{n}\right).$$

If we replace the definite integral in the equation

$$u(x) = f(x) + \int_a^b K(x, \xi) u(\xi) d\xi \tag{1}$$

---

* *Torino* I.

† Sur une nouvelle méthode pour la résolution du problème de Dirichlet, *Öfversigt af Kongl. Vetenskaps-Akademiens Förhandlingar*, vol. 57 (1900), p. 39. This is the Swedish academy at Stockholm. A second paper, complete in itself, and much more extensive, appears in *Acta Math.* vol. 27 (1903), p. 365.

‡ *Göttinger Nachrichten*, 1904, p. 49. This is the first of a series of papers in which important contributions were made to the theory. The plan of deducing the results as limiting cases of *algebraic* propositions as the number of variables becomes infinite is consistently carried through.

by the sum of which it is the limit, we get the equation

$$u(x) = f(x) + \sum_{j=1}^{n} K(x, x_j) u(x_j) \delta \qquad (2).$$

We shall see in a moment that this equation has in general one and only one solution, and we may expect that this solution, which we denote by $u_n(x)$, will have as its limit for $n = \infty$ the desired solution of (1).

Since equation (2) is to hold for all values of x in ab, it must in particular hold when $x = x_1, x_2, \dots x_n$. This gives us the system of n equations

$$-\sum_{j=1}^{n} K(x_i, x_j) u_n(x_j) \delta + u_n(x_i) = f(x_i) \quad (i = 1, 2, \dots n) \qquad (3).$$

These may be regarded as n non-homogeneous linear equations for determining the n unknowns $u_n(x_1), \dots u_n(x_n)$. When these have been determined, the value of $u_n(x)$ may be found by substituting these values in the second member of (2). Consequently, if the determinant

$$D_n = \begin{vmatrix} -\delta K(x_1, x_1) + 1 & -\delta K(x_1, x_2) & \dots & -\delta K(x_1, x_n) \\ -\delta K(x_2, x_1) & -\delta K(x_2, x_2) + 1 & \dots & -\delta K(x_2, x_n) \\ \dotfill & & & \\ \dotfill & & & \\ -\delta K(x_n, x_1) & -\delta K(x_n, x_2) & \dots & -\delta K(x_n, x_n) + 1 \end{vmatrix}$$

of the system (3) does not vanish, we see that, as was asserted above, there exists one and only one solution of (2). In fact, if we desire not $u_n(x)$ itself but merely its limit for $n = \infty$, we need not consider (2) at all, for this limit, which we assume to be continuous, is completely determined by the values of $u_n(x_i)$ obtained from (3).

The determinant D_n when expanded takes the form

$$D_n = 1 - \sum_{i=1}^{n} \delta K(x_i, x_i) + \frac{1}{2!} \sum_{i,j=1}^{n} \delta^2 \begin{vmatrix} K(x_i, x_i) & K(x_i, x_j) \\ K(x_j, x_i) & K(x_j, x_j) \end{vmatrix}$$

$$-\frac{1}{3!} \sum_{i,j,k=1}^{n} \delta^3 \begin{vmatrix} K(x_i, x_i) & K(x_i, x_j) & K(x_i, x_k) \\ K(x_j, x_i) & K(x_j, x_j) & K(x_j, x_k) \\ K(x_k, x_i) & K(x_k, x_j) & K(x_k, x_k) \end{vmatrix}$$

$$+ \dots + (-1)^n \delta^n \begin{vmatrix} K(x_1, x_1) \dots K(x_1, x_n) \\ \dotfill \\ \dotfill \\ K(x_n, x_1) \dots K(x_n, x_n) \end{vmatrix} {}^* .$$

---

* The factorials in the denominators of all but the last term are extremely important. They are due to the fact that in the following Σ every term is repeated the number of times indicated by the factorial in question. For the sake of simplicity, the notation has been changed in the last term, so that the factorial does not appear.

If we allow n to become infinite, we get in this expression a larger and larger number of terms, and the terms themselves vary and approach definite limits. We are thus led to the consideration of the infinite series

$$D = 1 - \int_a^b K(\xi_1, \xi_1)\, d\xi_1 + \frac{1}{2!} \int_a^b \int_a^b \begin{vmatrix} K(\xi_1, \xi_1) & K(\xi_1, \xi_2) \\ K(\xi_2, \xi_1) & K(\xi_2, \xi_2) \end{vmatrix} d\xi_1 d\xi_2 - \dots.$$

We shall prove in the next section that this series converges. Its value is called by Fredholm the *determinant* of the integral equation (1) or of the kernel K. We shall not stop to prove that it is the limit of D_n for $n = \infty$, although the method of deriving it makes this extremely plausible*.

Let us further consider the cofactor of the determinant D_n which corresponds to the νth row and the μth column. We readily see that, when $\mu \neq \nu$, this determinant may be written

$$D_n(x_\mu, x_\nu) = \delta \left\{ K(x_\mu, x_\nu) - \sum_{i=1}^n \delta \begin{vmatrix} K(x_\mu, x_\nu) & K(x_\mu, x_i) \\ K(x_i, x_\nu) & K(x_i, x_i) \end{vmatrix} \right.$$

$$+ \frac{1}{2!} \sum_{i,j=1}^n \delta^2 \begin{vmatrix} K(x_\mu, x_\nu) & K(x_\mu, x_i) & K(x_\mu, x_j) \\ K(x_i, x_\nu) & K(x_i, x_i) & K(x_i, x_j) \\ K(x_j, x_\nu) & K(x_j, x_i) & K(x_j, x_j) \end{vmatrix}$$

$$\left. - \dots + \frac{(-1)^{n-2}}{(n-2)!} \Sigma \delta^{n-2} \begin{vmatrix} (n-1)\text{-rowed} \\ \text{determinants} \end{vmatrix} \right\}.$$

If here we let n become infinite, and at the same time allow μ and ν to vary in such a way that $\lim (x_\mu, x_\nu) = (x, y)$, we get† as the limit of $\delta^{-1} D_n(x_\mu, x_\nu)$ the series

$$D(x, y) = K(x, y) - \int_a^b \begin{vmatrix} K(x, y) & K(x, \xi_1) \\ K(\xi_1, y) & K(\xi_1, \xi_1) \end{vmatrix} d\xi_1$$

$$+ \frac{1}{2!} \int_a^b \int_a^b \begin{vmatrix} K(x, y) & K(x, \xi_1) & K(x, \xi_2) \\ K(\xi_1, y) & K(\xi_1, \xi_1) & K(\xi_1, \xi_2) \\ K(\xi_2, y) & K(\xi_2, \xi_1) & K(\xi_2, \xi_2) \end{vmatrix} d\xi_1 d\xi_2 - \dots.$$

This series, the proof of whose convergence is reserved for the next section, we shall call the *adjoint* of the kernel K‡.

It should be noticed that we have not here considered the cofactors

---

* The proof here required is very similar to the one which we meet in the elements of analysis when we prove that the expansion of $(1 + 1/m)^m$ by the binomial theorem (m a positive integer) approaches as its limit for $m = \infty$ the familiar series for e. Both of these proofs, and many others of a similar character, may be easily carried through by means of a general theorem of Osgood, *Annals of Math.* ser. 2, vol. 3 (1902), p. 138, Ex. 8.

† This statement of course requires proof. Cf. the preceding foot-note.

‡ Fredholm uses the term *first minor*.

of D_n which correspond to the elements of its principal diagonal. These cofactors differ in form only slightly from the determinant D_n itself, and it is clear that we shall get as their limits when $n = \infty$ precisely the determinant D of the integral equation.

Let us now proceed to solve the system (3) of equations on the supposition that $D_n \neq 0$. We have by Cramer's formulae

$$u_n(x_\mu) = \frac{f(x_1) D_n(x_\mu, x_1) + f(x_2) D_n(x_\mu, x_2) + \dots + f(x_n) D_n(x_\mu, x_n)}{D_n}$$

$$(\mu = 1, 2, \dots n).$$

If we let n become infinite, and allow at the same time μ to vary in such a way that x_μ approaches x as a limit, we obtain as the limiting form of this formula, when we remember that

$$\lim D_n(x_\mu, x_\mu) = D,$$

$$u(x) = f(x) + \frac{1}{D} \int_a^b f(\xi) D(x, \xi) d\xi.$$

This is actually the solution of the integral equation (3) as found by Fredholm.

Instead of placing this solution on a firm foundation by justifying all the steps we have taken, we prefer to establish it with Fredholm *ab initio* in the next section. The method to be used will be suggested to us if we recall how Cramer's formulae for the solution of (3) are established; namely by multiplying the equations (3) by the cofactors $D_n(x_\mu, x_i)$ and adding them together. The essential point here is that this has the effect of eliminating all the unknowns except $u_n(x_\mu)$. This is due to the theorem which says that if we take the elements of the μth column of D_n and multiply them respectively into the cofactors of the elements of another column, the sum of the results thus obtained will be zero. This may be expressed by the formula

$$-\delta \sum_{i=1}^{n} K(x_i, x_\mu) D_n(x_\lambda, x_i) + D_n(x_\lambda, x_\mu) = 0 \qquad (\lambda \neq \mu).$$

If we here divide by $-\delta$ and then let n become infinite, at the same time allowing x_λ and x_μ to approach respectively the values x and y, we get as the limiting form of this formula

$$\int_a^b K(\xi, y) D(x, \xi) d\xi + K(x, y) D - D(x, y) = 0 \qquad (4).$$

This suggests that (4) is the essential instrument by which we shall solve the integral equation (1) in the next section. This formula must of course first be established more firmly than has yet been done.

By the side of this formula is another similar one* at which we arrive by starting from the fact that if in the determinant D_n we multiply the elements of the νth *row* by the cofactors of the corresponding elements of another row and add the results together, the sum is zero. This may be expressed by the formula

$$- \delta \sum_{i=1}^{n} K(x_\nu, x_i) D_n(x_i, x_\lambda) + D_n(x_\nu, x_\lambda) = 0.$$

By taking limits, as above, we are led to the second fundamental formula

$$\int_a^b K(x, \xi) D(\xi, y) \, d\xi + K(x, y) D - D(x, y) = 0 \qquad (5).$$

8. Fredholm's Solution. We begin by establishing a theorem concerning determinants due to Hadamard†. For this purpose consider the following simple geometrical fact.

If a parallelopiped has one vertex at the origin and the three adjacent vertices at the points (x_1, y_1, z_1), (x_2, y_2, z_2), (x_3, y_3, z_3), it is well known that its volume will be

$$\Delta = \begin{vmatrix} x_1 & y_1 & z_1 \\ x_2 & y_2 & z_2 \\ x_3 & y_3 & z_3 \end{vmatrix}.$$

Let us suppose that the three edges issuing from the origin are of unit length $$x_i^2 + y_i^2 + z_i^2 = 1 \qquad (i = 1, 2, 3) \qquad (1).$$
Then it is clear geometrically that the volume of the parallelopiped will be a maximum when the three edges in question are mutually perpendicular, in which case the volume is 1. Hence under the conditions (1) we see that $|\Delta| \leqq 1$. This suggests at once the following generalization :

LEMMA. *If the elements a_{ij} of the determinant*

$$\Delta = \begin{vmatrix} a_{11} & \dots & a_{1n} \\ \dots & & \dots \\ \dots & & \dots \\ a_{n1} & \dots & a_{nn} \end{vmatrix}$$

are real and satisfy the conditions

$$a_{i1}^2 + a_{i2}^2 + \dots + a_{in}^2 = 1 \qquad (i = 1, 2, \dots n) \quad (2),$$
then $$|\Delta| \leqq 1.$$

---

* This formula, or something equivalent to it, is necessary in showing that Cramer's formulae really satisfy the system of linear equations.

† *Bull. des sciences math. et astr.* 2nd ser. vol. 17 (1893), p. 240. Hadamard's method is purely algebraic. We follow a method of Wirtinger, *Monatshefte f. Math. u. Physik*, vol. 18 (1907), p. 158. Both authors consider also more general questions, in particular they consider the case in which the elements of the determinant are imaginary. Cf. also Fischer, *Archiv d. Math. u. Phys.* 3 ser., vol. 13 (1908), p. 32.

In order to prove this, let us attempt to find the maxima and minima of Δ under the conditions (2).

In the first place it is clear that Δ has both a finite maximum and a finite minimum under the conditions (2), since these conditions restrict the point $(a_{11}, a_{12}, \ldots a_{nn})$ in space of n^2 dimensions to a finite closed region, and Δ is a continuous function of these arguments. Moreover since Δ has continuous first partial derivatives with regard to these arguments, we may apply the ordinary method of the differential calculus for finding maxima and minima. Let as allow

$$a_{i1}, a_{i2}, \ldots a_{in}$$

to vary, leaving the other a's constant. We thus get from (2)

$$a_{i1}da_{i1} + a_{i2}da_{i2} + \ldots + a_{in}da_{in} = 0 \qquad (3);$$

and when we remember that $\partial\Delta/\partial a_{ij}$ is the cofactor A_{ij} of a_{ij} in Δ, we see that a necessary condition for a maximum or minimum is

$$A_{i1}da_{i1} + A_{i2}da_{i2} + \ldots + A_{in}da_{in} = 0 \qquad (4).$$

Moreover since, when Δ is a maximum or minimum, every set of values of the differentials which satisfy (3) also satisfy (4), the first members of (3) and (4) are proportional to each other, and we have

$$A_{ij} = \lambda_i a_{ij} \qquad (i, j = 1, 2, \ldots n).$$

Multiplying these equations by a_{ij} and adding, we get, on referring to (2),

$$\Delta = \lambda_i,$$

so that*

$$A_{ij} = \Delta a_{ij} \qquad (i, j = 1, 2, \ldots n) \qquad (5).$$

The determinant of the nth order of which A_{ij} is the general element has, as is well known, the value Δ^{n-1}. Forming this determinant from the values of A_{ij} in (5), we thus get

$$\Delta^{n-1} = \Delta^{n+1}.$$

Consequently, when Δ is a maximum or minimum,

$$\Delta = \pm 1.$$

The maximum value of Δ is therefore $+1$ and the minimum -1, and our lemma is proved.

We can now readily pass to a more general result. Let us suppose that the elements of Δ are still real but that conditions (2) are not fulfilled. Let us denote the value of the left-hand side of (2) by σ_i. If, for the moment, we rule out the possibility of all the elements of one

---

* Notice that, if $\Delta \neq 0$, this is precisely the condition that Δ be an orthogonal determinant.

row of Δ vanishing, these σ_i's are all positive, and we may consider the determinant

$$\frac{\Delta}{\sqrt{\sigma_1 \sigma_2 \ldots \sigma_n}} = \begin{vmatrix} \dfrac{a_{11}}{\sqrt{\sigma_1}} & \cdots & \dfrac{a_{1n}}{\sqrt{\sigma_1}} \\ \cdots\cdots\cdots\cdots \\ \cdots\cdots\cdots\cdots \\ \dfrac{a_{n1}}{\sqrt{\sigma_n}} & \cdots & \dfrac{a_{nn}}{\sqrt{\sigma_n}} \end{vmatrix}.$$

This determinant satisfies all the conditions of our lemma, and hence we have the inequality

$$|\Delta| \leq \sqrt{\sigma_1 \sigma_2 \ldots \sigma_n}.$$

If now we let M be a positive constant at least as great as any of the quantities $|a_{ij}|$, we see that

$$\sigma_i \leq nM^2$$

and thus we obtain the formula

$$|\Delta| \leq \sqrt{n^n}\, M^n.$$

This inequality obviously also holds in the case we have excluded in which all the elements in some row of Δ are zero ; and we have proved

HADAMARD'S THEOREM. *If the elements a_{ij} of the determinant*

$$\Delta = \begin{vmatrix} a_{11} & \cdots & a_{1n} \\ \cdots\cdots\cdots \\ \cdots\cdots\cdots \\ a_{n1} & \cdots & a_{nn} \end{vmatrix}$$

are real and satisfy the inequality

$$|a_{ij}| \leq M,$$

then $$|\Delta| \leq \sqrt{n^n}\, M^n.$$

We proceed now to Fredholm's solution of the integral equation

$$u(x) = f(x) + \int_a^b K(x, \xi)\, u(\xi)\, d\xi \tag{6}$$

in which we assume* that $K(x, \xi)$ is finite in S and that any discontinuities it may have are regularly distributed, while $f(x)$ is assumed to be continuous in I. Furthermore we will assume that $K(x, x)$ is integrable throughout the interval I.

---

* We shall not consider the case in which K fails to remain finite. This case is of considerable practical importance, and we refer the reader to Fredholm's, Hilbert's, and E. Schmidt's papers for various treatments of it.

We form the two series

$$D = 1 - \int_a^b K(\xi_1,\ \xi_1)\,d\xi_1 + \frac{1}{2!}\int_a^b\int_a^b \begin{vmatrix} K(\xi_1,\ \xi_1) & K(\xi_1,\ \xi_2) \\ K(\xi_2,\ \xi_1) & K(\xi_2,\ \xi_2) \end{vmatrix} d\xi_1 d\xi_2$$

$$- \frac{1}{3!}\int_a^b\int_a^b\int_a^b \begin{vmatrix} K(\xi_1,\ \xi_1) & K(\xi_1,\ \xi_2) & K(\xi_1,\ \xi_3) \\ K(\xi_2,\ \xi_1) & K(\xi_2,\ \xi_2) & K(\xi_2,\ \xi_3) \\ K(\xi_3,\ \xi_1) & K(\xi_3,\ \xi_2) & K(\xi_3,\ \xi_3) \end{vmatrix} d\xi_1 d\xi_2 d\xi_3 + \dots \quad (7),$$

$$D(x,y) = K(x,y) - \int_a^b \begin{vmatrix} K(x,y) & K(x,\ \xi_1) \\ K(\xi_1,y) & K(\xi_1,\ \xi_1) \end{vmatrix} d\xi_1$$

$$+ \frac{1}{2!}\int_a^b\int_a^b \begin{vmatrix} K(x,y) & K(x,\ \xi_1) & K(x,\ \xi_2) \\ K(\xi_1,y) & K(\xi_1,\ \xi_1) & K(\xi_1,\ \xi_2) \\ K(\xi_2,y) & K(\xi_2,\ \xi_1) & K(\xi_2,\ \xi_2) \end{vmatrix} d\xi_1 d\xi_2 - \dots \quad (8).$$

All of the multiple integrals which occur here obviously converge and may be evaluated as iterated integrals taken in any order.

We will now prove that the first of these series converges absolutely, and that the second converges absolutely and uniformly in S. As was already stated in the last section, we shall call the constant D the determinant of K, and the function $D(x,y)$ the adjoint of K^*.

Let M be the upper limit of $|K(x,y)|$ in S. Then each of the determinants which occur in (7) satisfies the conditions of Hadamard's Theorem, and the general term of (7)

$$\frac{(-1)^n}{n!}\int_a^b \dots \int_a^b \begin{vmatrix} K(\xi_1,\ \xi_1) & \dots & K(\xi_1,\ \xi_n) \\ \dots & & \dots \\ \dots & & \dots \\ K(\xi_n,\ \xi_1) & \dots & K(\xi_n,\ \xi_n) \end{vmatrix} d\xi_1 \dots d\xi_n$$

does not exceed in absolute value

$$C_n = \frac{\sqrt{n^n}\,M^n\,(b-a)^n}{n!}.$$

The series of which C_n is the general term converges, since

$$\frac{C_{n+1}}{C_n} = \frac{1}{\sqrt{n+1}}\sqrt{\left(1 + \frac{1}{n}\right)^n}\,M(b-a),$$

a quantity which obviously approaches zero as its limit when $n = \infty$. Consequently the series (7) converges absolutely.

Similarly the general term of (8) does not exceed in absolute value

$$B_n = \frac{\sqrt{n^n}\,M^n\,(b-a)^{n-1}}{(n-1)!};$$

and

$$\frac{B_{n+1}}{B_n} = \frac{\sqrt{n+1}}{n}\sqrt{\left(1 + \frac{1}{n}\right)^n}\,M(b-a).$$

---

* Strictly speaking, the determinant and adjoint *with regard to the region S*.

Since this last quantity approaches zero as its limit, the series of which B_n is the general term converges; and, this being a series of constant positive terms, the series (8) converges absolutely and uniformly in S.

It will, of course, be essential for us to discuss the question of the continuity or discontinuity of the adjoint function $D(x, y)^*$. For this purpose let us write the general term of (8) in the form

$$\frac{(-1)^n}{n!} K(x, y) \int_a^b \cdots \int_a^b \begin{vmatrix} K(\xi_1, \xi_1) \cdots K(\xi_1, \xi_n) \\ \cdots\cdots\cdots\cdots\cdots\cdots \\ \cdots\cdots\cdots\cdots\cdots\cdots \\ K(\xi_n, \xi_1) \cdots K(\xi_n, \xi_n) \end{vmatrix} d\xi_1 \cdots d\xi_n$$

$$+ \frac{(-1)^n}{n!} Q_n(x, y) \qquad (9),$$

where

$$Q_n(x, y) = \int_a^b \cdots \int_a^b \begin{vmatrix} 0 & K(x, \xi_1) & \cdots K(x, \xi_n) \\ K(\xi_1, y) & K(\xi_1, \xi_1) & \cdots K(\xi_1, \xi_n) \\ \cdots\cdots\cdots\cdots\cdots\cdots\cdots\cdots \\ \cdots\cdots\cdots\cdots\cdots\cdots\cdots\cdots \\ K(\xi_n, y) & K(\xi_n, \xi_1) & \cdots K(\xi_n, \xi_n) \end{vmatrix} d\xi_1 \cdots d\xi_n$$

$$(10).$$

Since the first term in (9) is the product of the general term of (7) by $K(x, y)$, we may write

$$D(x, y) = D \cdot K(x, y) + \sum_{n=1}^{\infty} \frac{(-1)^n}{n!} Q_n(x, y) \qquad (11).$$

If we expand the determinant in (10) according to the elements of its first row, we get

$$Q_n = \sum_{i=1}^{n} (-1)^i \int_a^b \cdots \int_a^b P_{ni} K(x, \xi_i) d\xi_1 \cdots d\xi_n \qquad (12),$$

where

$$P_{ni} = \begin{vmatrix} K(\xi_1, y) \cdots K(\xi_1, \xi_{i-1}) K(\xi_1, \xi_{i+1}) \cdots K(\xi_1, \xi_n) \\ \cdots\cdots\cdots\cdots\cdots\cdots\cdots\cdots\cdots\cdots \\ \cdots\cdots\cdots\cdots\cdots\cdots\cdots\cdots\cdots\cdots \\ K(\xi_n, y) \cdots K(\xi_n, \xi_{i-1}) K(\xi_n, \xi_{i+1}) \cdots K(\xi_n, \xi_n) \end{vmatrix}$$

Let us change the notation for the variables of integration by introducing in the ith integral in (12) ξ in place of ξ_i, ξ_i in place of ξ_{i+1}, ξ_{i+1} in place of ξ_{i+2}, etc., but leaving the variables $\xi_i, \ldots \xi_{i-1}$ unchanged in notation. If then in this ith integral we bring the ith

---

* If we wished to assume that $K(x, y)$ is continuous throughout S, we could easily infer from the continuity of the terms in (8), and from the uniform convergence of this series, that $D(x, y)$ is continuous in S.

row of the determinant into the first place, all the terms of (12) are seen to be equal, and we may write

$$Q_n(x, y) = -n \int_a^b \cdots \int_a^b K(x, \xi) P_n d\xi d\xi_1 \cdots d\xi_{n-1} \quad (13),$$

where

$$P_n = \begin{vmatrix} K(\xi, y) & K(\xi, \xi_1) & \cdots K(\xi, \xi_{n-1}) \\ K(\xi_1, y) & K(\xi_1, \xi_1) & \cdots K(\xi_1, \xi_{n-1}) \\ \cdots\cdots\cdots\cdots\cdots\cdots\cdots\cdots\cdots\cdots \\ \cdots\cdots\cdots\cdots\cdots\cdots\cdots\cdots\cdots\cdots \\ K(\xi_{n-1}, y) & K(\xi_{n-1}, \xi_1) & \cdots K(\xi_{n-1}, \xi_{n-1}) \end{vmatrix}.$$

We are now in a position to examine the question of the continuity of $D(x, y)$. For this purpose let us first show that the functions $Q_n(x, y)$ as defined by (10) are continuous throughout S. From either (10) or (13) it is clear by a reference to Theorem 1, § 1, that $Q_1(x, y)$ is continuous throughout S. Let us assume that $Q_1, \ldots Q_{n-1}$ are continuous there. If from this assumption we can infer that Q_n is continuous there, we shall have completed the proof that all Q's are continuous by the method of mathematical induction. We may write the second member of (13) in the form

$$-n \int_a^b \cdots \int_a^b K(x, \xi) K(\xi, y) \begin{vmatrix} K(\xi_1, \xi_1) & \cdots K(\xi_1, \xi_{n-1}) \\ \cdots\cdots\cdots\cdots\cdots\cdots\cdots \\ \cdots\cdots\cdots\cdots\cdots\cdots\cdots \\ K(\xi_{n-1}, \xi_1) & \cdots K(\xi_{n-1}, \xi_{n-1}) \end{vmatrix} d\xi \cdots d\xi_{n-1}$$

$$-n \int_a^b \cdots \int_a^b K(x, \xi) \begin{vmatrix} 0 & K(\xi, \xi_1) & \cdots K(\xi, \xi_{n-1}) \\ K(\xi_1, y) & K(\xi_1, \xi_1) & \cdots K(\xi_1, \xi_{n-1}) \\ \cdots\cdots\cdots\cdots\cdots\cdots\cdots\cdots\cdots \\ K(\xi_{n-1}, y) & K(\xi_{n-1}, \xi_1) & \cdots K(\xi_{n-1}, \xi_{n-1}) \end{vmatrix} d\xi \cdots d\xi_{n-1}.$$

The first of these integrals is merely a constant multiple of

$$\int_a^b K(x, \xi) K(\xi, y) d\xi,$$

and is therefore continuous by Theorem 1, § 1. The second may be written

$$-n \int_a^b K(x, \xi) Q_{n-1}(\xi, y) d\xi,$$

and is therefore continuous for the same reason.

Having thus seen that all the Q's are continuous, we infer that the function represented by the series in (11) is also continuous, since this series is uniformly convergent in S, being the difference between the uniformly convergent series (8) and the product of the series (7) by the finite function $K(x, y)$. We can now read off from (11) the finiteness and the nature of the discontinuities of $D(x, y)$.

The fundamental identity which connects $K(x, y)$, $D(x, y)$, and D can now be established. If we substitute the value of Q_n from (13) in the series in (11), we get the same series we should obtain by multiplying the series for $D(\xi, y)$ by $K(x, \xi)$ and integrating term by term. Since this integration is permissible on account of the uniform convergence of (8) and of the integrability of the value of this series and of its terms, we have thus established the formula

$$\sum_{n=1}^{\infty} \frac{(-1)^n}{n!} Q_n(x, y) = \int_a^b K(x, \xi) D(\xi, y) d\xi \qquad (14).$$

In precisely the same way, if we expand the determinant in (10) according to the elements of its first column, we find

$$\sum_{n=1}^{\infty} \frac{(-1)^n}{n!} Q_n(x, y) = \int_a^b D(x, \xi) K(\xi, y) d\xi \qquad (15).$$

By combining (14) and (15) with (11), we thus get the fundamental formula

$$D(x, y) - D . K(x, y) = \int_a^b K(x, \xi) D(\xi, y) d\xi$$
$$= \int_a^b D(x, \xi) K(\xi, y) d\xi \qquad (16).$$

This is the same as formulae (4) and (5) of the preceding section, which were there deduced without any attempt at accuracy.

We have thus proved

THEOREM 1. *Every function $K(x, y)$ which is finite in S and whose discontinuities are regularly distributed and for which $K(x, x)$ is integrable in I has a determinant D given by the absolutely convergent series (7) and an adjoint $D(x, y)$ given by the absolutely and uniformly convergent series (8). This adjoint function is finite in S and is continuous at every point where K is continuous*. The function K, its determinant and its adjoint satisfy the identity (16).*

If $D \neq 0$, we may divide (16) by $-D$, and if we let

$$k(x, y) = -\frac{D(x, y)}{D} \qquad (17),$$

* If $D=0$, $D(x, y)$ is continuous throughout S. If $D \neq 0$, $D(x, y)$ is discontinuous wherever $K(x, y)$ is discontinuous, and is discontinuous in such a way that $D(x, y) - D . K(x, y)$ is continuous.

the identity (16) reduces to formula (9), § 6.   Hence

THEOREM 2. *Every function $K(x, y)$ finite in S, whose disconti-nuities are regularly distributed, for which $K(x, x)$ is integrable in I, and whose determinant is different from zero has a reciprocal, which is given by formula* (17).

By a reference to Theorem 1, § 6, we get at once the further result

THEOREM 3. *If the kernel $K(x, y)$ of the equation* (6) *is finite in S, and any discontinuities it may have are regularly distributed, and $K(x, x)$ is integrable in I, and if its determinant D is not zero, and $f(x)$ is continuous in I, the equation* (6) *has one and only one continuous solution, and this solution is given by the formula*

$$u(x) = f(x) + \int_a^b \frac{D(x, \xi)}{D} f(\xi) \, d\xi.$$

In order to secure the existence of a determinant and an adjoint we have been obliged to assume that $K(x, x)$ is integrable, since otherwise even the terms of the series (7), (8) would be meaningless. Such a restriction on the kernel of equation (6) is, however, obviously immaterial since a change of definition of K on the line $x = y$ will clearly have no effect on the solutions of this equation. In order to cover all cases conveniently, we will lay down the following

DEFINITION *. *Let $K(x, y)$ be any function finite in S and whose discontinuities are regularly distributed, and let*

$$K_0(x, y) = \begin{cases} K(x, y) \text{ when } x \neq y, \\ 0 \quad \text{ when } x = y; \end{cases}$$

then the determinant D_0 and the adjoint $D_0(x, y)$ of K are called the modified determinant and the modified adjoint of K respectively.

Using this definition, we obtain at once

THEOREM 4. *If we drop the requirement that $K(x, x)$ be integrable in I, Theorem 3 remains true provided we use in place of D and $D(x, y)$ the modified determinant and the modified adjoint of K respectively.*

It is to be noticed that the modified determinant and the modified adjoint may exist when the determinant and adjoint do not. If, how-ever, the conditions of Theorem 1 are fulfilled, so that all four of these

* The idea involved in this definition was used by Hilbert (*Göttinger Nachrichten*, 1904, p. 82) for the purpose of treating a case in which K is not necessarily finite.

quantities exist, it is a matter of some interest to know the relation between the determinant and the modified determinant, and also between the adjoint and the modified adjoint. These relations are given by the formulae

$$D_0 = e^c D \qquad (18),$$

$$D_0(x, y) = \begin{cases} e^c D(x, y) & \text{when } x \neq y \\ e^c [D(x, y) - DK(x, y)] & \text{when } x = y \end{cases} \qquad (19),$$

where D and $D(x, y)$ are the determinant and adjoint of K, and D_0 and $D_0(x, y)$ the modified determinant and modified adjoint, and

$$c = \int_a^b K(\xi, \xi) \, d\xi \qquad (20).$$

The proof of (18) and the first half of (19) consists simply in multiplying the series for D_0 or $D_0(x, y)$ by the series

$$e^{-c} = 1 - \int_a^b K(\xi, \xi) \, d\xi + \frac{1}{2!} \left(\int_a^b K(\xi, \xi) \, d\xi \right)^2 - \ldots,$$

while the second half of (19) may readily be deduced either in the same way or from (16) in combination with (18) and the first half of (19). The details of these proofs we leave to the reader*.

The case considered by Liouville and Volterra in which the upper limits of integration are variable may readily be seen to come under the last theorem. We get this case, as we have seen, by supposing that $K(x, y) = 0$ when $y > x$. It is not hard to show that $D_0 = 1$, and that the series (8) for $D_0(x, y)$ reduces to the series (5) of § 6, so that the solution just given reduces in this case precisely to Volterra's solution.

In conclusion, we turn very briefly to the case $D = 0$. If (6) has a continuous solution $u(x)$, we shall have

$$u(\xi) = f(\xi) + \int_a^b K(\xi, \xi_1) u(\xi_1) \, d\xi_1.$$

Multiply this equation by $D(x, \xi)$ and integrate with regard to ξ from a to b. By reducing the result obtained by means of (16), we deduce

---

* The reader may also show that if $K(x, y)$ is finite in S, and any discontinuities it may have are regularly distributed, and its modified determinant $D_0 \neq 0$, it has a reciprocal which is given by the formula

$$k(x, y) = \begin{cases} -\dfrac{D_0(x, y)}{D_0} & (x \neq y), \\ -\dfrac{D_0(x, y)}{D_0} - K(x, y) & (x = y). \end{cases}$$

THEOREM 5*. *If the determinant of the integral equation* (6) *vanishes, while the other conditions of Theorem* 3 *are fulfilled, a necessary condition for the equation to have a solution continuous throughout I is*

$$\int_a^b D(x, \xi) f(\xi) d\xi = 0.$$

Apart from the special case in which $D(x, y) \equiv 0$, it will be seen that when $D = 0$ it is only for *special* functions $f(x)$ that the equation (6) can have a continuous solution.

Finally we note that if we drop the restriction that $K(x, x)$ be integrable, Theorem 5 remains true if we replace the determinant and the adjoint by the modified determinant and the modified adjoint.

9.   The Integral Equation with a Parameter.   For many purposes it is important to consider integral equations of the second kind whose kernel contains a parameter λ. In particular, the case in which this parameter comes in only as a factor is of prime importance. In this case the equation may be written

$$u(x) = f(x) + \lambda \int_a^b K(x, \xi) u(\xi) d\xi \dagger \qquad (1).$$

It is customary to speak of K as the kernel of this equation. It seems more consistent, however, to call λK the kernel, and this we shall do.

It will be assumed throughout this section that f is continuous throughout I, that K is finite in S, that any discontinuities K may have are regularly distributed, and that $K(x, x)$ is integrable throughout $I\ddagger$. While we still suppose the functions f and K to be real§, we shall allow the parameter λ to take on complex as well as real values. This makes the kernel λK no longer necessarily real, but this, as was remarked in the second footnote to § 5, will not in any way affect the developments of that section and the next, the main results of which read as follows when we replace K by λK:

---

* This theorem obviously corresponds to the fact that if the determinant of a system of linear equations vanishes, the equations have, in general, no solution.

† It may be noticed that if we let $\lambda = 1/\rho$, Liouville's original equation (5), § 4, has precisely this form.

‡ This last restriction need not be made for Theorems 1 and 2 below, while in the later theorems it might easily be avoided by using the modified determinant and adjoint.

§ It may readily be seen that the reality of K is not necessary for the proofs of Theorems 1 and 2. That it is not necessary in the case of the later theorems becomes evident if we make use of the more general form of Hadamard's Theorem which refers to determinants with complex elements.

THEOREM 1. *If K_1, K_2, ... are the functions obtained by iteration from K, then for any value of λ for which the series*

$$k(x, y ; \lambda) = -\lambda K_1(x, y) - \lambda^2 K_2(x, y) - \dots \qquad (2)$$

converges uniformly in (x, y) throughout S, the function $k(x, y ; \lambda)$ is the reciprocal of $\lambda K(x, y)$; and, for any such value of λ, the equation (1) has one and only one continuous solution, namely

$$u(x) = f(x) - \int_a^b k(x, \xi ; \lambda) f(\xi) d\xi \qquad (3),$$

a solution which is obviously real if λ is real.

THEOREM 2. *If M denotes the upper limit of $|K(x, y)|$ in S, the hypothesis of Theorem 1 will be fulfilled when*

$$|\lambda| < \frac{1}{M(b-a)} \qquad (4).$$

This hypothesis will be fulfilled for all values of λ if

$$K(x, y) = 0 \qquad \text{when } y > x \qquad (5).$$

Turning now to § 8, we first establish

THEOREM 3. *The two power-series*

$$D(\lambda) = 1 - \lambda \int_a^b K(\xi_1, \xi_1) d\xi_1$$
$$+ \frac{\lambda^2}{2!} \int_a^b \int_a^b \begin{vmatrix} K(\xi_1, \xi_1) & K(\xi_1, \xi_2) \\ K(\xi_2, \xi_1) & K(\xi_2, \xi_2) \end{vmatrix} d\xi_1 d\xi_2 - \dots \qquad (6),$$

$$D(x, y ; \lambda) = \lambda K(x, y) - \lambda^2 \int_a^b \begin{vmatrix} K(x, y) & K(x, \xi_1) \\ K(\xi_1, y) & K(\xi_1, \xi_1) \end{vmatrix} d\xi_1$$
$$+ \frac{\lambda^3}{2!} \int_a^b \int_a^b \begin{vmatrix} K(x, y) & K(x, \xi_1) & K(x, \xi_2) \\ K(\xi_1, y) & K(\xi_1, \xi_1) & K(\xi_1, \xi_2) \\ K(\xi_2, y) & K(\xi_2, \xi_1) & K(\xi_2, \xi_2) \end{vmatrix} d\xi_1 d\xi_2 - \dots (7),$$

converge for all values of λ, and the second converges uniformly in $(x, y ; \lambda)$ for (x, y) in S and λ restricted to any finite range of values. For this range of values of $(x, y ; \lambda)$, the function $D(x, y ; \lambda)$ represented by the second series is finite and has no discontinuities except where K is discontinuous.

These functions satisfy the identity

$$D(x, y ; \lambda) - \lambda D(\lambda) K(x, y) = \lambda \int_a^b K(x, \xi) D(\xi, y ; \lambda) d\xi$$
$$= \lambda \int_a^b K(\xi, y) D(x, \xi ; \lambda) d\xi \qquad (8).$$

If we restricted ourselves to real values of λ, this theorem would be hardly more than a special case of the results of § 8 obtained by replacing K by λK, the one new point being that the convergence of (7) is uniform in λ as well as in (x, y). This point, however, becomes at once obvious when we replace λ by a positive constant μ such that for the range of values of λ considered $|\lambda| \leqq \mu$.

In the case where λ is complex* the theorem cannot be obtained as a special case of § 8, since in that section we assumed that K was real, and λK, which now takes the place of K, is not real. The proof of the theorem is, however, immediate if we run through the developments of § 8 again replacing K by λK at every point. We leave it for the reader to do this.

We add, also without proof, the following theorem which the reader may establish by reasoning closely analogous to that used in § 8.

THEOREM 4. *The series obtained by differentiating (7) p times term by term with regard to λ converges uniformly in $(x, y; \lambda)$ for (x, y) in S and λ restricted to any finite range of values. The function*

$$\partial^p D(x, y; \lambda)/\partial \lambda^p$$

represented by this series is finite for this range of values and is discontinuous only where K is discontinuous.

We next establish a lemma concerning power-series which we shall find useful.

LEMMA†. *If in the power-series*

$$P(x, y; \lambda) = a_0(x, y) + a_1(x, y) \lambda + a_2(x, y) \lambda^2 + \dots \quad (9)$$

each of the coefficients $a_i(x, y)$ is finite in S, and if there exists a positive constant R such that when $|\lambda| < R$ the series (9) converges throughout S; and if the function P represented by the series is a finite function of $(x, y; \lambda)$ when (x, y) varies throughout S and λ varies on the circumference $|\lambda| = r$, where r is a positive constant less than R; then, if ρ is a positive constant less than r, the series (9) converges uniformly in $(x, y; \lambda)$ for (x, y) in S and $|\lambda| \leqq \rho$.

For denote by M the upper limit of $|P|$ when (x, y) is in S and $|\lambda| = r$. By a well-known theorem on power-series (cf. Forsyth, *Theory of Functions*, 2nd ed., p. 34)

$$M \geqq |a_n| r^n.$$

* No special consideration of this case would be necessary if Hadamard's Theorem had been established for determinants whose elements are complex.

† This lemma may be extended at once to the case in which the a's are functions of any number of variables, and S is replaced by any finite or infinite region.

Consequently, when (x, y) is in S and $|\lambda| \leq \rho$, the terms of (9) do not exceed in absolute value the corresponding terms of the series

$$M + M\frac{\rho}{r} + M\left(\frac{\rho}{r}\right)^2 + \dots,$$

and, this being a convergent series of positive constant terms, the uniform convergence of (9) is established.

The exceptional character of the case in which $D = 0$ was evident in § 8. We lay down the following

DEFINITION. *The roots of the integral analytic function $D(\lambda)$ shall be called simply the roots for the function $K(x, y)$*.*

We are already familiar with one case in which $D(\lambda)$ has no roots, namely the case in which $K(x, y)$ is zero when $y > x$. A simple case in which $D(\lambda)$ does have a root is when $K(x, y) = 1$. Here

$$D(\lambda) = 1 - \lambda(b - a),$$

so that $D(\lambda)$ has one and only one root, namely $\lambda = 1/(b - a)$. Other important cases in which $D(\lambda)$ has at least one root will be considered in § 11.

By the proof used for Theorem 2, § 8, we obtain

THEOREM 5. *Except when λ is a root for $K(x, y)$, the function $\lambda K(x, y)$ has a reciprocal $k(x, y; \lambda)$ given by the formula*

$$k(x, y; \lambda) = -\frac{D(x, y; \lambda)}{D(\lambda)} \qquad (10).$$

This reciprocal is therefore an analytic function of λ which has no singularities in the finite part of the λ-plane except perhaps poles at the roots of $D(\lambda)$. We will now show that at any such point k does have a pole, at least for some points (x, y) in S.

For this purpose we first note the formula

$$\int_a^b D(\xi, \xi; \lambda)d\xi = -\lambda\frac{dD(\lambda)}{d\lambda} \qquad (11),$$

whose truth follows at once from formulae (6), (7) †.

Let λ_1 be the root we wish to consider, and suppose that it is a root of the mth order of $D(\lambda)$, so that we may write

$$D(\lambda) = c_m(\lambda - \lambda_1)^m + c_{m+1}(\lambda - \lambda_1)^{m+1} + \dots \quad (m > 0, c_m \neq 0) \quad (12),$$

this equation being valid for all values of λ.

---

* We note that the roots of the modified determinant $D_0(\lambda)$ are the same as those of $D(\lambda)$, cf. formula (18), § 8. We may therefore compute the roots for $K(x, y)$ as the roots of $D_0(\lambda)$. We should so *define* them if $K(x, x)$ were not integrable.

† That $D(\xi, \xi; \lambda)$ is integrable in I is clear, since by (11), § 8, it differs by a continuous function from a constant multiple of $K(\xi, \xi)$. This same formula shows us that the series for $D(\xi, \xi; \lambda)$ may be integrated term by term.

On the other hand $D(x, y; \lambda)$, being analytic in λ for all finite points in the λ-plane, may be developed about $\lambda = \lambda_1$ in a power-series whose coefficients are functions of (x, y). Some of the first of these coefficients may be identically zero throughout S. They cannot, however, all be identically zero, for then, as we see from (7), we should have $K(x, y) \equiv 0$, and consequently $D(\lambda) \equiv 1$, and this is inconsistent with the hypothesis that $D(\lambda)$ has a root λ_1. We may therefore write

$$D(x, y; \lambda) = g_k(x, y)(\lambda - \lambda_1)^k + g_{k+1}(x, y)(\lambda - \lambda_1)^{k+1} + \ldots \quad (k \geqq 0, g_k \not\equiv 0)$$
(13).

The coefficients $g_k, g_{k+1}, \ldots$ differ merely by constant factors from successive derivatives of $D(x, y; \lambda)$ with regard to λ when $\lambda = \lambda_1$. From Theorem 4 we therefore see that all these g's are finite throughout S and are discontinuous only where K is discontinuous.

A reference to our Lemma now shows us that the series (13) converges uniformly in $(x, y; \lambda)$ when (x, y) varies throughout S and λ varies over any finite region in the λ-plane. We may then substitute the series (12) and (13) in (11), and integrate term by term on the left, thus getting

$$\sum_{i=k}^{\infty} (\lambda - \lambda_1)^i \int_a^b g_i(\xi, \xi) \, d\xi = \lambda \sum_{i=m}^{\infty} ic_i (\lambda - \lambda_1)^{i-1}.$$

If here we write $\lambda = \lambda_1 + (\lambda - \lambda_1)$, we see that the second member of this equation may be developed in a series proceeding according to positive integral powers of $\lambda - \lambda_1$, and beginning with the term in $(\lambda - \lambda_1)^m$ whose coefficient is not zero, since $\lambda_1 \neq 0$, $c_m \neq 0$. Since the left-hand side is also a power-series in $\lambda - \lambda_1$, it follows that it must also begin with the same term, so that $k \leqq m - 1$, that is

$$k < m \tag{14}.$$

Since $k(x, y; \lambda)$ is the negative of the ratio of (13) to (12), we thus get

THEOREM 6. *If λ_1 is a root for $K(x, y)$, and $k(x, y; \lambda)$ the reciprocal of $\lambda K(x, y)$, then, at least for some points in S, $k(x, y; \lambda)$ has a pole at the point $\lambda = \lambda_1$.*

Instead of considering the development of k about a root of $D(\lambda)$, let us now consider its development about the point $\lambda = 0$. This is certainly not a root of $D(\lambda)$, as we see from the form of the series for $D(\lambda)$. Consequently the function $k(x, y; \lambda)$ is analytic in λ at the point $\lambda = 0$, and can be developed about this point in a power-series which converges in a circle described about the origin as centre and reaching out to the point in the λ-plane nearest to the origin which represents a root of $D(\lambda)$. It is possible that for some points (x, y) this series may converge outside of this circle, but it cannot do so for all points of S. This development must be precisely the series (2), since we have seen that that series represents $k(x, y; \lambda)$ when $|\lambda|$ is sufficiently small.

By means of our Lemma we can go a step further. For, if we denote by r a positive constant smaller than the absolute value of any of the roots of $D(\lambda)$, the function $D(\lambda)$ is continuous and different from zero when $|\lambda| = r$. We know also from Theorem 3 that $D(x, y; \lambda)$ is finite when $|\lambda| = r$ and (x, y) is in S. Consequently, by (10), the same will be true of $k(x, y; \lambda)$. We may then infer by means of our Lemma the uniform convergence of (2), and we thus get

THEOREM 7. *If λ_1 is the root for $K(x, y)$ of smallest absolute value, the series (2) converges uniformly in $(x, y; \lambda)$ when $|\lambda| \leqq \rho < |\lambda_1|$, and (x, y) is in S, to the function $k(x, y; \lambda)$ reciprocal to $\lambda K(x, y)$.*

From Theorem 5 and Theorem 1, § 6, we infer at once

THEOREM 8. *If λ is not one of the roots for K, equation (1) has one and only one solution continuous in x, namely*

$$u(x, \lambda) = f(x) + \int_a^b \frac{D(x, \xi; \lambda)}{D(\lambda)} f(\xi)\, d\xi.$$

It will readily be seen that $u(x, \lambda)$ is a function analytic at all finite points of the λ-plane except at the roots of $D(\lambda)$ where it can have no other singularities than poles.

10. The Fundamental Theorem concerning Homogeneous Integral Equations, with some Applications. We consider in this section the equation

$$u(x) = \lambda \int_a^b K(x, \xi)\, u(\xi)\, d\xi \qquad (1)$$

where, as before, we assume that K is finite in S, that any discontinuities it may have are regularly distributed, and that $K(x, x)$ is integrable in I^*.

From Theorem 8, § 9, we see that, when λ is not a root for K, the only continuous solution of (1) is the obvious solution $u = 0$. It remains merely to consider the case in which λ is a root of $D(\lambda)$. We will prove that in this case (1) always has a continuous solution which does not vanish identically †.

---

* As in § 9, this last restriction might be avoided by using the modified determinant and adjoint.

† This fundamental theorem was first established by Fredholm. We follow Kneser, *Rendiconti* of Palermo, vol. 22 (1906), p. 233.

Let λ_1 be the root in question, and develop the determinant and the adjoint of λK into the series (12) and (13) of § 9. These series we substitute in the identity (8), § 9, and, since the second converges uniformly, we obtain, after dividing by $(\lambda - \lambda_1)^k$, the formula

$$\sum_{i=k}^{\infty} g_i(x, y)(\lambda - \lambda_1)^{i-k} - \lambda K(x, y) \sum_{i=m}^{\infty} c_i(\lambda - \lambda_1)^{i-k}$$
$$= \lambda \sum_{i=k}^{\infty} (\lambda - \lambda_1)^{i-k} \int_a^b K(x, \xi) g_i(\xi, y) d\xi \quad (2).$$

If here we let $\lambda = \lambda_1$, we see, on referring to the inequality $k < m$ (cf. (14), § 9), that all the terms reduce to zero except the first term in the first series and the first term in the last series. We thus get

$$g_k(x, y) = \lambda_1 \int_a^b K(x, \xi) g_k(\xi, y) d\xi \quad (3).$$

Thus we see that whatever constant value in the interval I we may assign to y, the function $g_k(x, y)$ is a solution of (1) when $\lambda = \lambda_1$. Moreover, since we saw in § 9 that g_k is finite in S and that any discontinuities it has are regularly distributed, we see, by referring to Theorem 1, § 1, that the second member of (2) is a continuous function of (x, y) throughout S. Hence g_k is a continuous solution of (1). Finally, we know that g_k does not vanish identically. Thus we have proved

THEOREM 1. *A necessary and sufficient condition that equation* (1) *have a continuous solution which is not identically zero is that* λ *be a root for* K.

DEFINITION 1. *By a principal solution for* $K(x, y)$ *is understood a continuous solution of the homogeneous equation* (1) *which is not identically zero* *.

The last theorem established shows that to every root of $D(\lambda)$ correspond one or more principal solutions. In fact it is easily seen that to a root of $D(\lambda)$ will always correspond an infinite number of such solutions. For if $u_1(x)$ is a first principal solution corresponding to the root λ_1, then $c_1 u_1(x)$, where c_1 is any constant different from zero, will clearly also be a principal solution corresponding to λ_1 †.

---

* A more complete term here would be principal solution in x (Nulllösung in x); by a principal solution in y being understood a continuous solution of

$$u(y) = \lambda \int_a^b K(x, y) u(x) dx$$

which does not vanish identically.

† Notice that the principal solution $g_k(x, y)$ obtained in the proof of Theorem 1 also contains a parameter, y.

If, besides the solutions $c_1 u_1(x)$, the equation (1) has, when $\lambda = \lambda_1$, other continuous solutions, let $u_2(x)$ be such a one. Then

$$c_1 u_1(x) + c_2 u_2(x)$$

will clearly be a continuous solution whatever the values of the constants c_1 and c_2 may be. Proceeding in this way, we see that *unless the equation* (1) *has when* $\lambda = \lambda_1$ *an infinite number of linearly independent continuous solutions* the general continuous solution of (1) for $\lambda = \lambda_1$ may be written

$$c_1 u_1(x) + c_2 u_2(x) + \ldots + c_n u_n(x),$$

where $u_1, \ldots u_n$ are linearly independent continuous solutions of (1) for $\lambda = \lambda_1$ and $c_1, \ldots c_n$ are arbitrary constants. The functions $u_1, \ldots u_n$ may then be called a fundamental system of solutions of (1) when $\lambda = \lambda_1$*; and the number n may be called the *index* of the root λ_1, a term not to be confounded with the *multiplicity* of this root. We give the formal definition of both of these terms:

DEFINITION 2. *The number of linearly independent principal solutions for $K(x, y)$ which correspond to a root λ_1 for K is called the index of λ_1; the number of times λ_1 occurs as a root of the analytic function $D(\lambda)$* (*the determinant of λK*) *is called the multiplicity of λ_1.*

From analogy with the theory of linear algebraic equations we should expect that the determination of the index of a root for K would require the introduction of series which correspond to the second minors, third minors, etc., of the determinant of the system or linear equations just as the series for $D(x, y; \lambda)$ corresponds to the first minors of this determinant. Such series were, in fact, introduced by Fredholm, and form an essential part of the article already cited. By means of them he was able to determine completely the index of any root of $D(\lambda)$, and among other things he established the interesting fact that *the index of a root of $D(\lambda)$ can never exceed its multiplicity.* In particular, since the multiplicity is necessarily finite, it follows that the index is also finite. For the discussion of these questions, we refer the reader to Fredholm's article. In § 12 we shall give a different proof, due to E. Schmidt, that the index of a *real* root of $D(\lambda)$ is finite.

We conclude this section with two fairly obvious applications of the theory of the homogeneous equation, first to the theory of reciprocal

* All this is in perfect analogy with the theory of homogeneous linear algebraic equations. Cf., for instance, the author's *Introduction to Higher Algebra* (Macmillan, 1907), p. 49.

functions, and secondly to the theory of the non-homogeneous equation.

THEOREM 2. *If $K(x, y)$ is finite in S and any discontinuities it may have are regularly distributed, and $K(x, x)$ is integrable in I, a necessary and sufficient condition that K have a reciprocal is that the determinant of K do not vanish.*

That this is a sufficient condition was already proved in Theorem 2, § 8. To prove it necessary, suppose the determinant were zero. Then the equation (1) has an infinite number of continuous solutions, and this is seen by Theorem 1, § 6, to be impossible if K has a reciprocal.

THEOREM 3. *If λ is a root for $K(x, y)$, the equation* *

$$u(x) = f(x) + \lambda \int_a^b K(x, \xi) u(\xi) d\xi \qquad (4)$$

has either no continuous solution or an infinite number of continuous solutions.

For it is clear that by adding to a first continuous solution of (4) any continuous solution of the homogeneous equation (1), we get another continuous solution of (4).

In fact it is readily seen that the *general* solution of (4) will be obtained by adding to a particular solution of (4) the general solution of (1). Cf. the corresponding well-known fact for non-homogeneous linear *differential* equations.

11. Symmetric Kernels. A function $K(x, y)$ is said to be symmetric if $K(x, y) \equiv K(y, x)$. Integral equations whose kernels are symmetric not only play, as has been shown by Hilbert, a very important part in the applications, but their theory may be used (cf. the papers of Hilbert and Schmidt†) as a foundation for the theory of integral equations whose kernel is not symmetric. We assume throughout this section that K is finite in S, and that its discontinuities are regularly distributed.

---

* We assume, of course, that $f(x)$ is continuous in I, and that K satisfies the conditions stated at the beginning of this section.

† A series of papers by Hilbert in the *Göttinger Nachrichten*, beginning in 1904, entitled "Grundzüge einer allgemeinen Theorie der linearen Integralgleichungen"; and E. Schmidt, *Math. Ann.* vol. 63 (1907), p. 433. The greater part of this last paper originally appeared in 1905 as a Dr. dissertation. A continuation, referring to non-linear integral equations, appeared in *Math. Ann.* vol. 65 (1908), p. 370.

THEOREM 1. *If $K(x, y)$ is symmetric and does not vanish at all points of S where it is continuous, then all of the iterated functions $K_2(x, y)$, $K_3(x, y)$, ... are symmetric and none of them is identically zero.*

For let $K_n(x, y)$ be the first of these functions which is not symmetric. Then

$$K_n(x, y) = \int_a^b K_{n-1}(x, \xi) K_1(\xi, y) \, d\xi \qquad (1).$$

This, by the symmetry of K_1 and K_{n-1}, reduces to

$$\int_a^b K_1(y, \xi) K_{n-1}(\xi, x) \, d\xi,$$

which is precisely $K_n(y, x)$. Thus K_n is symmetric.

On the other hand, if some of the iterated functions vanish identically, let $K_{n-1}(x, y)$ be the first one to do so. We see by (1) that $K_n(x, y)$ also vanishes identically. One of the two integers $n-1$ and n is even. Calling this even integer $2m$, we have

$$0 \equiv K_{2m}(x, y) \equiv \int_a^b K_m(x, \xi) K_m(\xi, y) \, d\xi.$$

Accordingly, owing to the symmetry of K_m,

$$0 \equiv K_{2m}(x, x) \equiv \int_a^b [K_m(x, \xi)]^2 \, d\xi.$$

This, however, is possible only if K_m vanishes at all points of S where it is continuous. Since, as we saw in § 6, K_2, K_3, ... are continuous throughout S, we are thus led to a contradiction, and our theorem is proved.

We now come to a fundamental existence theorem first established by Hilbert, and which we shall prove by a method due to Kneser*.

THEOREM 2. *For every symmetric function $K(x, y)$ which does not vanish at every point where it is continuous, there is at least one root.*

This theorem will be established (cf. Theorem 7, § 9) if we can show that the series (2) of § 9 is not uniformly convergent in (x, y) for every value of λ. Let us then assume that it is uniformly convergent throughout S for every value of λ. The series

$$\lambda^2 K_2(x, x) + \lambda^3 K_3(x, x) + \dots$$

is then, for every value of λ, uniformly convergent throughout I. It can

---

* *Loc. cit.* p. 236.

therefore, since its terms are continuous, be integrated term by term, and, if we let

$$U_n = \int_a^b K_n(x, x)\, dx \qquad (2),$$

the series $U_2\lambda^2 + U_3\lambda^3 + \ldots$

is convergent for all values of λ. Since this is a power-series, it is necessarily absolutely convergent, and hence a series formed from part of its terms is convergent. Thus

$$U_2\lambda^2 + U_4\lambda^4 + U_6\lambda^6 + \ldots \qquad (3)$$

converges for all values of λ.

If, now, we remember that K is symmetric, we get, by formula (4), § 6,

$$U_{n+m} = \int_a^b \int_a^b K_n(x, \xi)\, K_m(x, \xi)\, d\xi dx,$$

and in particular

$$U_{2n} = \int_a^b \int_a^b [K_n(x, \xi)]^2\, d\xi dx \qquad (4).$$

Hence by expanding the obvious inequality

$$\int_a^b \int_a^b [pK_{n+1}(x, \xi) + qK_{n-1}(x, \xi)]^2\, d\xi dx \geqq 0,$$

where p and q are arbitrary real parameters, we get

$$p^2 U_{2n+2} + 2pq U_{2n} + q^2 U_{2n-2} \geqq 0;$$

that is, the first member of this inequality is a positive definite quadratic form in (p, q). Consequently

$$U_{2n+2} U_{2n-2} - U^2_{2n} \geqq 0;$$

or, since by (4), in combination with Theorem 1, the U's with even subscripts are positive (not zero),

$$\frac{U_{2n+2}}{U_{2n}} \geqq \frac{U_{2n}}{U_{2n-2}} \qquad (5).$$

Now in the series (3), the ratio of two successive terms is

$$\frac{U_{2n+2}}{U_{2n}} \lambda^2,$$

which, by a successive application of (5), we see is not less than

$$\frac{U_4}{U_2} \lambda^2.$$

If then we let $\lambda = \sqrt{U_2/U_4}$, we see that the terms of (2) do not decrease as we go out in the series, and consequently the series cannot converge. Since this is a contradiction, our theorem is proved.

Our proof establishes at once the

COROLLARY. *Under the hypothesis of Theorem 2, there is at least one root for $K(x, y)$ which in absolute value does not exceed $\sqrt{U_2/U_4}$, where the U's are defined by* (2).

Let us now suppose that for the symmetric function $K(x, y)$ there are *two* roots, λ_1 and λ_2, and let us denote by $u_1(x)$ and $u_2(x)$ principal solutions for K which correspond to λ_1 and λ_2 respectively. Then

$$u_1(x) = \lambda_1 \int_a^b K(x, \xi)\, u_1(\xi)\, d\xi,$$

$$u_2(x) = \lambda_2 \int_a^b K(x, \xi)\, u_2(\xi)\, d\xi.$$

Let us multiply the first of these equations by $\lambda_2 u_2(x)$, the second by $\lambda_1 u_1(x)$ and subtract. This gives

$$(\lambda_2 - \lambda_1)\, u_1(x)\, u_2(x) = \lambda_1\lambda_2 \int_a^b K(x, \xi) \left[u_1(\xi)\, u_2(x) - u_2(\xi)\, u_1(x) \right] d\xi,$$

and from this follows

$$(\lambda_2 - \lambda_1) \int_a^b u_1(x)\, u_2(x)\, dx$$
$$= \lambda_1\lambda_2 \int_a^b \int_a^b \left[K(x, \xi)\, u_1(\xi)\, u_2(x) - K(\xi, x)\, u_1(x)\, u_2(\xi) \right] d\xi dx.$$

By interchanging the variables of integration in the second half of this double integral, we see that the second member of this equation reduces to zero, and, since by hypothesis $\lambda_2 \neq \lambda_1$, we get the result

THEOREM 3. *If $u_1(x)$ and $u_2(x)$ are principal solutions for a symmetric function $K(x, y)$ which correspond to two distinct roots, then*

$$\int_a^b u_1(x)\, u_2(x)\, dx = 0.$$

The method we have just used is essentially that invented by Poisson in the analogous case of linear differential equations of the second order, and we will follow out the analogy still further by proving, also by Poisson's method, as Schmidt has done,

THEOREM 4*. *For the symmetric function $K(x, y)$, there can be only real roots.*

Since the coefficients of the power-series $D(\lambda)$ are real, imaginary

---

* This theorem was first proved by a wholly different method by Hilbert as an extension of a well-known theorem concerning symmetrical determinants. Cf. Weber, *Algebra*, 2nd. Ed. vol. 1, p. 309.

roots necessarily occur in conjugate imaginary pairs. If possible let $\mu + vi$ and $\mu - vi$ be two imaginary roots. Let $v(x) + iw(x)$ be a principal solution corresponding to $\mu + vi$, so that

$$v(x) + iw(x) = (\mu + vi) \int_a^b K(x, \xi)[v(\xi) + iw(\xi)] \, d\xi.$$

If in this equation we separate real and imaginary parts, and then recombine the resulting equations, we see that $v(x) - iw(x)$ is a principal solution corresponding to $\mu - vi$. Consequently by Theorem 3

$$\int_a^b \{[v(x)]^2 + [w(x)]^2\} \, dx = 0.$$

This, however, is impossible, since v and w cannot both vanish identically, as otherwise $v + iw$ would not be a principal solution. Thus our theorem is proved.

We saw in § 9 that the roots for $K(x, y)$ are the poles of the function $k(x, y; \lambda)$ reciprocal to $\lambda K(x, y)$. In the case we are now considering, in which $K(x, y)$ is symmetric, we will show that these poles are of the first order; or, to use the notation of § 9, that for every root λ_1, $m = k + 1$. Suppose this were not the case, so that $m > k + 1$. Then from formula (2) of § 10 we should not only get, by letting $\lambda = \lambda_1$,

$$g_k(x, y) = \lambda_1 \int_a^b K(x, \xi) g_k(\xi, y) \, d\xi \qquad (6),$$

as in § 10, but also, by first differentiating with regard to λ and then letting $\lambda = \lambda_1$,

$$g_{k+1}(x, y) = \int_a^b K(x, \xi) g_k(\xi, y) \, d\xi + \lambda_1 \int_a^b K(x, \xi) g_{k+1}(\xi, y) \, d\xi.$$

This may be reduced by (6) to the form

$$g_{k+1}(x, y) = \frac{g_k(x, y)}{\lambda_1} + \lambda_1 \int_a^b K(x, \xi) g_{k+1}(\xi, y) \, d\xi \qquad (7).$$

We now multiply (7) by $g_k(x, y)$, (6) by $g_{k+1}(x, y)$, subtract (6) from (7), and integrate the resulting equation with regard to x from a to b. This gives

$$0 = \frac{1}{\lambda_1} \int_a^b [g_k(x, y)]^2 \, dx$$

$$+ \lambda_1 \int_a^b \int_a^b K(x, \xi)[g_k(x, y) g_{k+1}(\xi, y) - g_{k+1}(x, y) g_k(\xi, y)] \, d\xi \, dx.$$

The double integral is seen, precisely as in a similar case in the proof of Theorem **3** above, to have the value zero, and we thus get the formula

$$\int_a^b [g_k(x,y)]^2\, dx = 0,$$

this equality holding for all values of y. This, however, is impossible, since $g_k(x, y)$ is not identically zero. Thus we have proved

THEOREM 5*. *If $K(x, y)$ is symmetric, and $k(x, y; \lambda)$ is the reciprocal of $\lambda K(x, y)$, then k has no singularities in the finite part of the λ-plane other than poles of the first order.*

It should be noticed that this theorem does not by any means assert that $D(\lambda)$ has only simple roots.

The results we have obtained in this section for symmetric kernels may be extended to certain more general cases by the following device due to Goursat†:

Form the function

$$K'(x, y) = \sqrt{p(x)\, p(y)\, q(x)\, q(y)}\; \kappa(x, y)$$

where κ is symmetric and does not vanish at every point where it is continuous, while p and q are continuous and positive (not zero) throughout I. The function K' then satisfies all the conditions imposed on K in Theorems 2, 4, 5. Now let

$$r(x) = \sqrt{\frac{p(x)}{q(x)}}$$

and form the function

$$K(x, y) = \frac{r(x)}{r(y)}\, K'(x, y) = p(x)\, q(y)\, \kappa(x, y).$$

Let us denote by $k'(x, y; \lambda)$ the reciprocal of $\lambda K'(x, y)$. Then the reciprocal of $\lambda K(x, y)$ will, by the foot-note to the definition of reciprocal functions in § 6, be

$$\frac{r(x)}{r(y)}\, k'(x, y; \lambda).$$

---

* Cf. the proof of this theorem in a somewhat more general case by Boggio, Paris *C. R.* Oct. 14, 1907. The case considered by Boggio is less general than the result of Goursat to be given in a moment. We mention in passing that the present theorem is the extension to the case of an infinite number of variables of part of the well-known algebraic theorem which says that the elementary divisors of the determinant D_n of § 7, when this determinant is symmetric and K is replaced by λK, are all of the first degree. Cf., for instance, the writer's *Introduction to Higher Algebra*, p. 305.

† Paris *C. R.* Feb. 17, 1908.

Now the roots for a function are the poles of the reciprocal of λ times this function. Hence the roots for $K(x, y)$ are the poles of the function last written, and since these are the same as the poles of $k'(x, y; \lambda)$, which in turn are the roots for $K'(x, y)$, we see, by Theorem 2, that there is necessarily at least one root for $K(x, y)$, and, by Theorem 4, that these roots are necessarily real. Finally, by Theorem 5, the reciprocal of $\lambda K(x, y)$ has no finite poles of order higher than the first. Thus we get the result, which includes Theorems 2, 4, 5 as special cases,

THEOREM 6. *If* $\kappa(x, y)$ *is symmetric and does not vanish at all points where it is continuous, and if* $p(x)$ *and* $q(x)$ *are continuous throughout* I *and do not vanish there*, then for the function*

$$K(x, y) = p(x) q(y) \kappa(x, y)$$

there is at least one root, all such roots are real, and the function reciprocal to $\lambda K(x, y)$ *has no finite poles of order higher than the first.*

12. Orthogonal Functions. We begin by laying down the following

DEFINITION 1. *The functions* $u_1(x)$, $u_2(x)$, *...,* *finite or infinite in number, are said to be orthogonal to each other in the interval* I *if*

(a) *they are real and continuous in* I;

(b) *no one of them is identically zero in* I;

(c) *every pair of them,* $u_i(x)$ *and* $u_j(x)$, *satisfy the relation*

$$\int_a^b u_i(\xi) u_j(\xi) \, d\xi = 0.$$

The connection between the subject of orthogonal functions and the theory of homogeneous integral equations with a symmetric kernel is evident from Theorem 3, § 11. On the other hand we recall that in other branches of mathematics, in particular in the developments of arbitrary functions which occur in mathematical physics, systems of orthogonal functions play an important part. Thus, for instance, in the subject of Fourier's Series we have to deal with the system of functions

$$\begin{cases} 1 & \cos x & \cos 2x & \cos 3x \dots \\ & \sin x & \sin 2x & \sin 3x \dots \end{cases}$$

orthogonal in the interval $0 \leqq x \leqq 2\pi$. Or again, the Legendre's Polynomials

$$P_0(x), \ P_1(x), \ P_2(x), \ ...,$$

---

* We are clearly justified in dropping the restriction that p and q be positive since a change of sign of κ does not affect our result.

in terms of which arbitrary functions may be developed in the interval $-1 \leqq x \leqq +1$, are orthogonal in this interval. On the other hand the Bessel's Function

$$J_0(a_1 x),\ J_0(a_2 x),\ \ldots,$$

in terms of which arbitrary functions may be developed in the interval $0 \leqq x \leqq 1$, where $a_1,\ a_2,\ \ldots$ are the roots of the transcendental equation $J_0(a) = 0$, are *not* orthogonal in this interval, but become so if they are all multiplied by the factor $\sqrt{x}$.

This subject of orthogonal functions is therefore one which has long been of importance for its own sake*. We shall first develop this subject independently, and then make some applications of the results obtained to the theory of integral equations†.

THEOREM 1. *If $u_1(x),\ \ldots u_n(x)$ are orthogonal in I, they are necessarily linearly independent there.*

For, if possible, let there be a relation

$$c_1 u_1(x) + \ldots + c_n u_n(x) = 0$$

where at least one of the constants c, say c_ν, is not zero. Multiplying this equation by $u_\nu(x)$ and integrating from a to b, we get

$$c_\nu \int_a^b [u_\nu(x)]^2\, dx = 0,$$

which is impossible.

THEOREM 2. *If $u_1(x),\ \ldots u_n(x)$ are real and continuous throughout I, and are linearly independent, there exist n linear combinations with constant real coefficients of these functions which are orthogonal in I.*

Let us begin by assuming that this theorem is true in the case of $n-1$ functions, so that we can get linear combinations with real constant coefficients $v_1(x),\ \ldots v_{n-1}(x)$ of the functions $u_1(x),\ \ldots u_{n-1}(x)$ which are orthogonal in I. It remains merely to show that the real constants c may be so determined that the function

$$v_n(x) = c_1 v_1(x) + \ldots + c_{n-1} v_{n-1}(x) + u_n(x)$$

is orthogonal to $v_1(x),\ \ldots v_{n-1}(x)$. Let k be any one of the integers $1, 2, \ldots n-1$. If we multiply the last equation by $v_k(x)$ and integrate from a to b, we get

$$\int_a^b v_n(x)\, v_k(x)\, dx = c_k \int_a^b [v_k(x)]^2\, dx + \int_a^b u_n(x)\, v_k(x)\, dx.$$

---

* So far as the writer knows, the *term* orthogonal was first used in this sense by F. Klein in a course of lectures on the differential equations of mathematical physics, delivered in Göttingen in the summer of 1889.

† Cf. Gram, *Crelle*, vol. 94 (1883), p. 41; and E. Schmidt, *Math. Ann.* vol. 63 (1907), p. 443.

If, then, we let

$$c_k = -\frac{\displaystyle\int_a^b u_n(x)\, v_k(x)\, dx}{\displaystyle\int_a^b [v_k(x)]^2\, dx} \qquad (k = 1, 2, \ldots n-1),$$

the function $v_n(x)$ will be orthogonal to $v_1(x), \ldots v_{n-1}(x)$.

In order that the proof of our theorem by mathematical induction be complete, it is merely necessary to notice that the theorem is true (though trivial) in the case of a single function.

We saw above that when we have an infinite set of orthogonal functions $u_1(x), u_2(x), \ldots$ the problem of developing an arbitrary function $f(x)$ in a series of the form

$$f(x) = c_1 u_1(x) + c_2 u_2(x) \ldots \tag{1}$$

frequently presents itself, the series to hold in the interval I in which the functions are orthogonal. Without attempting to consider this question, we merely give the familiar *formal* determination of the coefficients, a determination which is completely justifiable if we know that $f(x)$ can be developed into a series of this form which is uniformly convergent, or, more generally, which after multiplication by any continuous function can be integrated term by term.

This formal determination of the coefficients consists in multiplying (1) by $u_i(x)$ and integrating term by term from a to b. All the terms of the series but one then drop out on account of the orthogonality of the functions, and we get

$$c_i = \frac{\displaystyle\int_a^b f(x)\, u_i(x)\, dx}{\displaystyle\int_a^b [u_i(x)]^2\, dx} \tag{2}.$$

The ordinary formulae for the coefficients of a Fourier's Series or of a series in terms of Legendre's Polynomials or of Bessel's Functions are of course only special cases of (2).

By multiplying the functions $u_1, u_2, \ldots$ by suitable real constants we can obviously make the denominators in (2) take on any positive values we please. If, in particular, we make them all take on the value 1, we say that the functions are *normalized*.

DEFINITION 2. *A set of functions orthogonal in I are said to be normalized if the integral of the square of each one extended over I is 1.*

A much more elementary problem than the problem of development we have just touched upon is the problem of getting an approximate representation of a function $f(x)$ in the interval I by means of a given finite set of orthogonal functions $u_1(x), \ldots u_k(x)$. We will suppose $f(x)$ to be finite in I and to have at most a finite number of discontinuities there. We wish to determine the constants $c_1, \ldots c_k$ in such a way that the function

$$F(x) = c_1 u_1(x) + \ldots + c_k u_k(x)$$

gives the best approximate representation of $f(x)$ in I in the sense of the method of least squares. That is, we wish to determine the constants $c_1, \ldots c_k$ so that

$$J = \int_a^b [f(x) - F(x)]^2 \, dx$$

shall be a minimum. We have

$$\frac{\partial J}{\partial c_i} = -2 \frac{\partial}{\partial c_i} \int_a^b f(x) F(x) \, dx + \frac{\partial}{\partial c_i} \int_a^b [F(x)]^2 \, dx.$$

Now since

$$\int_a^b [F(x)]^2 \, dx = \int_a^b \{c_1^2 [u_1(x)]^2 + \ldots + c_k^2 [u_k(x)]^2\} \, dx,$$

we get

$$\frac{\partial J}{\partial c_i} = -2 \int_a^b f(x) u_i(x) \, dx + 2c_i \int_a^b [u_i(x)]^2 \, dx.$$

Equating this to zero, we get as the desired values of $c_1, \ldots c_k$ precisely the quantities (2).

Since $\partial^2 J / \partial c_i^2 > 0$, $\partial^2 J / \partial c_i \partial c_j = 0$, we see that formulae (2) really correspond to a minimum of J, and we have thus proved

THEOREM 3. *If $u_1(x), \ldots u_k(x)$ are orthogonal in I and $f(x)$ is finite in I and has at most a finite number of discontinuities, a necessary and sufficient condition that the finite series*

$$c_1 u_1(x) + \ldots + c_k u_k(x)$$

give the best representation of $f(x)$ in I in the sense of the method of least squares is that the c's be determined by (2).

For the sake of getting simpler formulae, let us suppose that the functions $u_i(x)$ are normalized. Then the minimum value of the integral J is seen, by making use of the values (2), to be

$$\int_a^b \left[f(x) - \sum_{i=1}^n u_i(x) \int_a^b f(\xi) u_i(\xi) \, d\xi \right]^2 dx$$

$$= \int_a^b [f(\xi)]^2 \, d\xi - \sum_{i=1}^n \left(\int_a^b f(\xi) u_i(\xi) \, d\xi \right)^2 \tag{3}.$$

Since the left-hand side of this equation is positive or zero, the same is true of the right-hand side, and we thus get the important inequality

$$\sum_{i=1}^{n} \left(\int_{a}^{b} f(\xi) u_i(\xi) d\xi \right)^2 \leq \int_{a}^{b} [f(\xi)]^2 d\xi * \qquad (4).$$

As an application of this inequality, we will establish the result, already stated without proof in § 10,

THEOREM 4. *If $K(x, y)$ is finite in S and its discontinuities are regularly distributed, then every real root for $K(x, y)$ has a finite index.*

For suppose that to the real root λ_1 there correspond n linearly independent real principal solutions for $K(x, y)$†. From these by means of Theorem 2 we can form n orthogonal principal solutions. Let us call these functions, after they have been normalized, $u_1(x), \ldots u_n(x)$. We will take these for the functions u of (4), and for $f(\xi)$ we will take $K(x, \xi)$, which, for any fixed value of x, is finite in I and has only a finite number of discontinuities there. When we remember that

$$u_i(x) = \lambda_1 \int_{a}^{b} K(x, \xi) u_i(\xi) d\xi \qquad (i = 1, 2, \ldots n),$$

the inequality (4) reduces to

$$\frac{1}{\lambda_1^2} \sum_{i=1}^{n} [u_i(x)]^2 \leq \int_{a}^{b} [K(x, \xi)]^2 d\xi.$$

If we integrate this with regard to x from a to b, we get on the left, when we remember that the u's are normalized, n/λ_1^2; and thus finally

$$n \leq \lambda_1^2 \int_{a}^{b} \int_{a}^{b} [K(x, \xi)]^2 d\xi dx \qquad (5).$$

This inequality gives us an upper limit for the number n of linearly independent real principal solutions corresponding to λ_1. It is easily seen, however, that the number of linearly independent complex principal solutions cannot be greater than the number of linearly independent real principal solutions. Thus our theorem is proved.

---

* Formula (3) is called by E. Schmidt *Bessel's Identity*, and (4) *Bessel's Inequality*, because in the *Astr. Nachr.* vol. 6 (1828), p. 333, Bessel had in the special case in which the u's are trigonometric functions considered a problem analogous to the one here treated, in which, however, no integrals but only finite sums present themselves. The first to consider precisely the problem here treated both for trigonometric functions and for Legendre's polynomials seems to have been Plarr, Paris *C. R.* vol. 44 (1857), p. 985.

† We do not here exclude the possibility of still other linearly independent principal solutions.

The inequality (5) may also be regarded as giving us a lower limit for the absolute value of the real root λ_1. By letting $n = 1$, we thus get a lower limit for the absolute value of any real root. Hence we see that in the case of a symmetrical kernel, or in the more general case of Theorem 6, § 11, no root can be in absolute value less than

$$\frac{1}{\sqrt{\int_a^b \int_a^b [K(x, \xi)]^2 \, d\xi \, dx}}.$$

In these cases we may therefore replace Theorem 2, § 6, by the following more far-reaching result:

THEOREM 5. *If the finite function K whose discontinuities are regularly distributed satisfies the conditions of Theorem 6, § 11, and*

$$\int_a^b \int_a^b [K(x, \xi)]^2 \, d\xi \, dx < 1,$$

the series (5) of § 6 converges absolutely and uniformly.

In the special case in which $K(x, y)$ is symmetric, the roots and the real principal solutions for K are also known as the *characteristic numbers* and the *characteristic functions** of K respectively. We saw in § 11 that these characteristic numbers are real, and that two characteristic functions corresponding to different characteristic numbers are orthogonal to each other. Moreover, by Theorem 4, we see that each characteristic number has a finite index. For each characteristic number whose index n is greater than 1, we can pick out a set of n orthogonal characteristic functions on which every other characteristic function corresponding to this characteristic number will then be linearly dependent. We thus get characteristic functions $u_1(x)$, $u_2(x)$, ..., finite or infinite in number as the case may be, orthogonal to each other in I, and such that every characteristic function of K is linearly dependent upon a finite number of them. Such a system we speak of as a *complete* orthogonal system of characteristic functions. We may, of course, normalize it if we choose.

A problem of fundamental importance is: under what conditions can a function $f(x)$ be developed in a series whose terms are constant multiples of the elements of this complete orthogonal system of characteristic functions, the coefficients of the series being determined by formula (2)? For such treatment as has, up to the present time, been given of this problem we refer to the papers already cited by Hilbert

---

* *Eigenwerte* and *Eigenfunktionen*. It is only when K is symmetric that the definition here given of these terms is correct. For the case of an unsymmetric kernel K, cf. the definition given by E. Schmidt, *loc. cit.* p. 461.

and Schmidt*. We confine ourselves to a single important, though very special, case, namely, that in which the function $f(x)$ to be developed is the symmetric kernel $K(x, y)$ itself, and even here we do not attempt to treat the problem completely.

Let us suppose that the complete orthogonal system of characteristic functions $u_1(x)$, $u_2(x)$, ... has been normalized, and denote by λ_1, λ_2, ... the characteristic numbers corresponding to them †. Formula (2), when we let $f(x) = K(x, y)$, now reduces to

$$c_\nu = \int_a^b K(x, y) u_\nu(x) dx = \frac{u_\nu(y)}{\lambda_\nu}.$$

We are thus led, by this purely formal work, to inquire whether the series

$$\frac{u_1(x) u_1(y)}{\lambda_1} + \frac{u_2(x) u_2(y)}{\lambda_2} + \dots \qquad (6)$$

really converges and represents the function $K(x, y)$. We prove here, following the method of Schmidt, the theorem due to Hilbert,

THEOREM 6. *If the symmetric function $K(x, y)$ is finite in S and its discontinuities are regularly distributed, and if $u_1(x)$, $u_2(x)$, ... is a complete normalized orthogonal system of characteristic functions of K, and λ_1, λ_2, ... are the corresponding characteristic numbers; then if the series (6) converges uniformly throughout S, it represents the function $K(x, y)$ at every point where this function is continuous.*

Since the series (6) converges uniformly and its terms are continuous, the function represented by (6) is continuous throughout S. It is also obviously symmetric. Consequently the function

$$Q(x, \xi) = K(x, \xi) - \Sigma \frac{u_i(x) u_i(\xi)}{\lambda_i} \qquad (7)$$

is symmetric and finite in S, and its discontinuities are regularly distributed. Let us assume that the theorem is false. Then Q does not vanish at every point where it is continuous; so that, by Theorem 2, § 11, it has at least one characteristic number c. Let $\psi(x)$ be a characteristic function of Q corresponding to c, so that

$$\psi(\xi) = c \int_a^b Q(\xi, \xi_1) \psi(\xi_1) d\xi_1 \qquad (8).$$

* Cf. also Kneser, *Math. Ann.* vol. 63 (1907), p. 477, where application is made to the problem of developing an arbitrary function according to the solutions of a linear differential equation of the second order. This application had already been made by Hilbert, though much less successfully.
† It should be noticed that the λ's are not necessarily all distinct.

If we multiply (7) by $u_i(\xi)\,d\xi$ and integrate, we get, when we remember that the u's are normalized,

$$\int_a^b Q(x,\,\xi)\,u_i(\xi)\,d\xi = \int_a^b K(x,\,\xi)\,u_i(\xi)\,d\xi - \frac{u_i(x)}{\lambda_i} = 0 \quad (9).$$

From (8) and (9) we deduce

$$\int_a^b \psi(\xi)\,u_i(\xi)\,d\xi = c\int_a^b\int_a^b Q(\xi,\,\xi_1)\,u_i(\xi)\,\psi(\xi_1)\,d\xi_1 d\xi = 0 \,(10).$$

Multiplying (7) by $\psi(\xi)\,d\xi$ and integrating, and reducing by means of (10), we find

$$\int_a^b Q(x,\,\xi)\,\psi(\xi)\,d\xi = \int_a^b K(x,\,\xi)\,\psi(\xi)\,d\xi.$$

The first member of this equation is, by (8), simply $\psi(x)/c$. Hence

$$\psi(x) = c\int_a^b K(x,\,\xi)\,\psi(\xi)\,d\xi.$$

Since $\psi(x)$ is continuous and does not vanish identically in I, this shows that $\psi(x)$ is a characteristic function of K, so that it must be linearly dependent on a finite number of the u's. This, however, is impossible, by Theorem 1, since, by (10), ψ is orthogonal to all the u's. Thus the theorem is proved, since the assumption that it is not true has led us to a contradiction.

The condition of uniform convergence of (6) is, of course, fulfilled if the series has only a finite number of terms, that is if K has only a finite number of characteristic numbers. In this case our theorem tells us that K is equal, wherever it is continuous, to the finite series (6), which is continuous throughout S. Any discontinuities which K has must therefore be *removable discontinuities*, that is, such that K becomes continuous throughout S if its definition *at its points of discontinuity* be suitably changed.

Conversely, if we can find a finite sum of the form

$$f_1(x)\,\phi_1(y) + f_2(x)\,\phi_2(y) + \ldots + f_n(x)\,\phi_n(y) \qquad (11)$$

where the f's and ϕ's are continuous in I, such that this sum is equal to K at all points of S where K is continuous, it is readily proved that K can have only a finite number of characteristic numbers. For, in the first place, in computing these characteristic numbers we may obviously use (11) in our kernel in place of $K(x, y)$. Now if we substitute (11) for K in the series for $D(\lambda)$ (formula (6), § 9), it will be seen that all the determinants of order higher than n in this series have the value zero. Consequently $D(\lambda)$ is a polynomial of at most the nth degree in λ, and cannot, therefore, have more than a finite number of roots. Thus we have proved

THEOREM 7. *If the symmetric function $K(x, y)$ is finite in S and its discontinuities are regularly distributed, a necessary and sufficient condition that K have only a finite number of characteristic numbers is that there exist a sum of the form* (11), *where the f's and ϕ's are continuous in I, which is equal to K at all points of S where K is continuous.*

In particular, if the symmetric function $K(x, y)$ is finite in S and has discontinuities *which are not all removable* and which are regularly distributed, it will surely have an infinite number of characteristic numbers.

13. The Integral Equation of the First Kind whose Kernel is Finite. In this section and the next we take up briefly the subject of integral equations of the first kind. We shall be concerned, in the main, with the equation

$$f(x) = \int_a^x K(x, \xi) u(\xi) d\xi \tag{1},$$

which has been so extensively treated by Volterra* that Picard has proposed to call it Volterra's Equation.

Let us assume that K is continuous throughout the triangle T, and that it has a derivative

$$K_1(x, y) = \frac{\partial K}{\partial x},$$

finite in T and whose discontinuities are regularly distributed. In order that (1) have a solution $u(x)$ continuous in I, it is evidently necessary that $f(x)$ be continuous in I and that $f(a) = 0$. Differentiating (1), we get

$$f'(x) = K(x, x) u(x) + \int_a^x K_1(x, \xi) u(\xi) d\xi \tag{2}.$$

From this we deduce at once, as a further necessary condition, that $f(x)$ have a derivative $f'(x)$ continuous throughout I. In order that (2) be an equation of the second kind satisfying the restrictions we have imposed in §§ 5, 6, it is now sufficient to impose on K the further restriction that $K(x, x)$ do not vanish at any point of I. Making this assumption, we can infer at once that (1) cannot have more than one continuous solution, and, if it has any such solution, this will be the continuous solution of (2).

* See the papers cited in § 6. Cf. also Holmgren, *Atti of the Turin Academy*, vol. 35 (1900), p. 570; and T. Labesco, *Journal de Mathématique*, 6th ser. vol. 4 (1908), p. 125.

We may also, if we impose the conditions just stated, work back from (2) to (1). For equation (2) may be written

$$\frac{d}{dx}\left[f(x) - \int_a^x K(x,\xi)\, u(\xi)\, d\xi \right] = 0.$$

Accordingly the continuous solution of (2) also satisfies the equation

$$f(x) - \int_a^x K(x,\xi)\, u(\xi)\, d\xi + \text{const.},$$

and by taking the limit as x approaches a, we see that this constant has the value zero. Thus we have proved

THEOREM 1*. *If $K(x,\xi)$ is continuous in T and has a derivative $K_1 = \partial K/\partial x$ finite in T and whose discontinuities are regularly distributed, and if $K(x,x)$ does not vanish at any point of I, a necessary and sufficient condition that* (1) *have a continuous solution is that $f(x)$ and its derivative $f'(x)$ be continuous in I and $f(a) = 0$. If these conditions are fulfilled,* (1) *has only one continuous solution, namely the continuous solution of the equation of the second kind* (2)†.

Without considering exhaustively the case in which $K(x,x)$ vanishes at one or more points in I, we wish to examine it sufficiently to show that it is really an exceptional case.

Let us first suppose that $K(x,x)$ vanishes identically in I. In this case (2) reduces to

$$f'(x) = \int_a^x K_1(x,\xi)\, u(\xi)\, d\xi \tag{3}.$$

Assuming that not only K but also K_1 is continuous in T, and that $K_1(x,x)$ does not vanish at any point of I, we see that equation (3) comes under the case covered by Theorem 1 provided that K_1 has a finite derivative whose discontinuities are regularly distributed. More-

* Cf. *Lincei* I. See also *Torino* I, and for a slightly earlier but less complete discussion, Le Roux, *Annales de l'École normale supérieure*, 3rd ser. vol. 12 (1895), p. 243.

† Volterra has also indicated a second method for reducing the equation (1) to an equation of the second kind. This consists in performing an integration by parts. If we let

$$K_2(x,y) = \frac{\partial K(x,y)}{\partial y}, \qquad U(x) = \int_a^x u(\xi)\, d\xi,$$

we get from (1) by an integration by parts

$$f(x) = K(x,x)\, U(x) - \int_a^x K_2(x,\xi)\, U(\xi)\, d\xi \tag{2'}.$$

We leave it for the reader to formulate conditions under which we can get the general continuous solution of (1) by differentiating the continuous solution of (2').

over we see as before that any continuous solution of (3) is also a solution of (1); and we get

THEOREM 2. *If $K(x, \xi)$ and $K_1(x, \xi) = \partial K/\partial x$ are continuous in T, and $K_{11}(x, \xi) = \partial^2 K/\partial x^2$ is finite in T and its discontinuities are regularly distributed, and $K(x, x) \equiv 0$ while $K_1(x, x)$ does not vanish at any point of I, then a necessary and sufficient condition that (1) have a continuous solution is that $f(x)$, $f'(x)$, $f''(x)$ be continuous in I and that $f(a) = f'(a) = 0$. If these conditions are fulfilled, the equation (1) has only one continuous solution, which may be found by solving (3).*

We leave it for the reader to push the results of this theorem farther by replacing the condition that $K_1(x, x)$ shall not vanish at any point of I by the condition $K_1(x, x) \equiv 0$.

The case in which $K(x, x)$ vanishes at a finite number of points in I has been discussed at length by Volterra*. We content ourselves with treating a simple example in which the results are fairly typical of the general case.

Consider the equation

$$0 = \int_0^x (a\xi + \beta x)\, u(\xi)\, d\xi \qquad (a + \beta \neq 0) \quad (4).$$

Here the kernel $a\xi + \beta x$ reduces when $\xi = x$ to $(a + \beta)\, x$, and thus vanishes when and only when $x = 0$. By differentiating (4) twice we see that any continuous solution which it may have has, except perhaps when $x = 0$, a continuous derivative $u'(x)$; and we obtain for $u(x)$ in this way the differential equation

$$(a + \beta)\, x\, u'(x) + (a + 2\beta)\, u(x) = 0 \qquad (5).$$

The general solution of this equation is

$$u(x) = cx^{-\frac{\beta}{a+\beta} - 1} \qquad (6).$$

If $c \neq 0$, this is continuous when and only when $\beta/(a + \beta) \leqq -1$; that is, when

$$0 > \frac{a + \beta}{\beta} \geqq -1,$$

or, more simply, when

$$-1 > \frac{a}{\beta} \geqq -2 \qquad (7).$$

If this condition is satisfied, the function (6) is readily seen to be a solution of (4), and since (6) contains an arbitrary constant, we have proved

---

* *Torino* III, IV. See also the papers by Holmgren and Labesco cited at the beginning of this section.

THEOREM 3. *Equation* (4) *has one and only one continuous solution (namely $u = 0$) except when condition* (7) *is fulfilled, and in which case it has an infinite number of such solutions.*

We leave it for the reader to extend this result to the more general equation

$$f(x) = \int_0^x (\alpha\xi + \beta x)\, u\,(\xi)\, d\xi \qquad (8),$$

where $f(x), f'(x), f''(x)$ are continuous in I, and $f(0) = f'(0) = 0$.

We may add that the method we have just used enables us to find not merely the continuous solutions of (4) or (8) but also, in some cases in which (7) is not fulfilled, discontinuous but integrable solutions. Thus, when $\alpha/\beta < -2$, the function (6) is integrable and is a solution of (4).

We turn now to the more general integral equation of the first kind

$$f(x) = \int_a^b K\,(x,\ \xi)\, u\,(\xi)\, d\xi \qquad (9).$$

Volterra's equation is the special case of this in which $K(x, y)$ vanishes when $y > x$. If we look more closely, we see that in the case covered by Theorem 1 the kernel $K(x, y)$ has a finite jump at every point of the line $x = y$, passing from the value $K(x, x)$ to the value 0 as we cross this line. The cases which presented more difficulty were those in which there is either no discontinuity along this line, or those in which there are a finite number of points where there is no discontinuity.

This suggests that we consider in place of the line $y = x$ the curve $y = \phi(x)$ (where we will assume $\phi(x)$ to be continuous and to have a continuous derivative $\phi'(x)$ in I), which we will suppose crosses the square S from the lower left-hand to the upper right-hand corner, so that $\phi(a) = a$, $\phi(b) = b$, and when

$$a < x < b, \quad a < \phi(x) < b.$$

Along this curve we suppose the function $K(x, y)$ to have a finite jump; that is, if (x_0, y_0) is any point of this curve at which

$$y_0 \neq a, \quad y_0 \neq b,$$

we assume that the two limits

$$K(x_0,\ y_0 - 0) \quad \text{and} \quad K(x_0,\ y_0 + 0)$$

exist. The difference of these is what we call the magnitude of the jump at (x_0, y_0), and we write

$$J(x_0) = K\,(x_0,\ y_0 - 0) - K\,(x_0,\ y_0 + 0).$$

We will suppose this function $J(x)$ to be continuous in I; and also that K is finite in S and continuous except along the curve $y = \phi(x)$, and that it has a derivative

$$K_1(x, y) = \partial K/\partial x$$

finite in S and whose discontinuities are regularly distributed.

Let us now write (9) in the form

$$f(x) = \int_a^{\phi(x)} K(x, \xi) u(\xi) d\xi + \int_{\phi(x)}^b K(x, \xi) u(\xi) d\xi.$$

By differentiating we get

$$f'(x) = J(x) \phi'(x) u(\phi(x)) + \int_a^b K_1(x, \xi) u(\xi) d\xi \qquad (10).$$

If neither J nor ϕ' vanishes in I, this equation reduces by means of the transformation $z = \phi(x)$, whose inverse we will write $x = \psi(z)$, to an integral equation of the second kind whose kernel is

$$\frac{-K_1(\psi(z), \xi)}{J(\psi(z)) \phi'(\psi(z))}.$$

If the determinant (or modified determinant) of this function is not zero, equation (10) has a continuous solution provided that $f'(x)$ be continuous. We cannot, however, infer from this that (9) also has a continuous solution. The continuous solution of (10) satisfies, as we readily see, an equation of the form

$$f(x) + \lambda = \int_a^b K(x, \xi) u(\xi) d\xi \qquad (11),$$

and satisfies it, of course, only for one value of λ. Thus we get the result:

THEOREM 4. *If the following conditions of continuity are satisfied*

(1) $K(x, y)$ *is finite in S, and, except on the curve $y = \phi(x)$, continuous there;*

(2) $\phi(x)$ *satisfies the conditions*

$$\phi(a) = a, \quad \phi(b) = b,$$

is continuous and has a continuous derivative in I, and $\phi'(x)$ does not vanish in I; the inverse of ϕ we denote by ψ;

(3) *At every point of I, except the ends, the two limits*

$$K(x, \phi(x) - 0) \quad and \quad K(x, \phi(x) + 0)$$

exist, and their difference

$$J(x) = K(x, \phi(x) - 0) - K(x, \phi(x) + 0)$$

is continuous and does not vanish in I;

(4) $K(x, y)$ *has a derivative*

$$K_1(x, y) = \partial K/\partial x$$

finite in S and whose discontinuities are regularly distributed;

(5) $f(x)$ *is continuous and has a continuous derivative in* I;

then if the determinant, or modified determinant, of the function

$$\frac{-K_1(\psi(z), \xi)}{J(\psi(z))\,\phi'(\psi(z))}$$

is not zero, there will be one and only one value of the parameter λ *for which the equation* (11) *has a continuous solution, and this solution will be the continuous solution of* (10).

The reader will find no difficulty in extending this theorem so as to cover cases in which $J(x) \equiv 0$.

14. Equations of the First Kind whose Kernel or whose Interval is not Finite. We pass now with Volterra* to the case in which the kernel K of the equation

$$f(x) = \int_a^x K(x, \xi)\, u(\xi)\, d\xi \tag{1}$$

has the form

$$K(x, \xi) = \frac{G(x, \xi)}{(x - \xi)^\lambda} \qquad (0 < \lambda < 1),$$

where G is continuous in T. Precisely as in the case of Abel's equation, of which (1) is an immediate generalization, we see that (1) cannot have a continuous solution unless $f(x)$ is continuous throughout I and $f(a) = 0$.

Assuming that (1) has a continuous solution, we may obtain it as follows: Multiply (1) by $(z - x)^{\lambda-1}\, dx$, where

$$a \leqq z \leqq b,$$

and integrate the resulting equation, getting

$$\int_a^z \frac{f(x)\, dx}{(z - x)^{1-\lambda}} = \int_a^z \frac{1}{(z - x)^{1-\lambda}} \int_a^x \frac{G(x, \xi)}{(x - \xi)^\lambda}\, u(\xi)\, d\xi\, dx \tag{2}.$$

The second member of this formula reduces by Dirichlet's Extended Formula (cf. § 1) to

$$\int_a^z u(\xi) \int_\xi^z \frac{G(x, \xi)}{(z - x)^{1-\lambda}(x - \xi)^\lambda}\, dx\, d\xi.$$

* Cf. *Torino* II.

Accordingly if we write

$$F(z) = \int_a^z \frac{f(x)}{(z-x)^{1-\lambda}}\, dx$$
$$L(z, \xi) = \int_\xi^z \frac{G(x, \xi)}{(z-x)^{1-\lambda}(x-\xi)^\lambda}\, dx \qquad (z > \xi) \qquad\qquad (3),$$

equation (2) takes the form

$$F(z) = \int_a^z L(z, \xi)\, u(\xi)\, d\xi \qquad\qquad (4).$$

We will now show that the kernel L of this equation is continuous in T, except on the line $z = \xi$ where it is not yet defined. For this purpose introduce y as variable of integration in place of x by means of the formula

$$y = \frac{x - \xi}{z - \xi}.$$

We thus get, when $z > \xi$,

$$L(z, \xi) = \int_0^1 \frac{G[(z-\xi)y + \xi, \xi]}{(1-y)^{1-\lambda} y^\lambda}\, dy \qquad\qquad (5).$$

Since this integral remains convergent when we replace G by the upper limit of its absolute value, it follows that it is uniformly convergent in T, and, since G is continuous, L is also continuous wherever it is defined.

In order next to see whether L approaches a limit as the point (z, ξ) approaches a point (c, c) on the hypotenuse of T, we apply the law of the mean for integrals to the original definition of L, getting

$$L(z, \xi) = G(s, \xi) \int_\xi^z \frac{dx}{(z-x)^{1-\lambda}(x-\xi)^\lambda} = \frac{\pi}{\sin \lambda\pi} G(s, \xi) \qquad (\xi < s < z).$$

Consequently

$$\lim_{\substack{z = c \\ \xi = c}} L(z, \xi) = \frac{\pi}{\sin \lambda\pi} G(c, c) \qquad\qquad (6).$$

Taking this limiting value as the *definition* of $L(c, c)$, we see that L is continuous throughout T, and that if we demand that $G(x, x)$ shall not vanish at any point of I, it follows that $L(x, x)$ does not vanish at any point of I.

In order then that the equation (4) come under the case covered by Theorem 1, § 13, it remains merely to impose on G restrictions which will make $L(z, \xi)$ have a partial derivative with regard to z finite in S and whose discontinuities are regularly distributed. In order to avoid all complications, we will assume that $G(z, \xi)$ has a partial derivative

$$G_1(z, \xi) = \partial G / \partial z,$$

B.

5

which is continuous in T. If, then, we differentiate (5) with regard to z under the integral sign, we get

$$\int_0^1 \left(\frac{y}{1-y}\right)^{1-\lambda} G_1\left[(z-\xi)y+\xi, \xi\right] dy.$$

Since this integral is obviously uniformly convergent, we see, by the ordinary test for differentiating an infinite integral under the sign of integration, that it represents

$$L_1(z, \xi) = \partial L/\partial z,$$

except perhaps on the hypotenuse $z = \xi$ of T where formula (5) was not valid. From the expression just obtained for L_1 we see that L_1 is finite in T, and from the uniform convergence of this integral, that L_1 is continuous in T except perhaps on the hypotenuse $z = \xi$.

It is now possible to apply Theorem 1 of § 13 to the equation (4). In order that this equation have a continuous solution it is therefore necessary and sufficient that $F'(z)$ be continuous and have a continuous derivative in I, and that $F(a) = 0$. The first and last of these conditions will be fulfilled if $f(x)$ is continuous in I, as we see by a reference to Theorem 2, § 1. We have thus obtained as an additional necessary condition for (1) to have a continuous solution that $F'(z)$ have a continuous derivative.

It remains to show that, if all the conditions we have mentioned are fulfilled, the continuous solution of (4) satisfies (1). For this purpose differentiate (4) with regard to z and then, after multiplying by $(x-z)^{-\lambda}dz$, where

$$a \leqq x \leqq b,$$

integrate from a to x, getting

$$\int_a^x \frac{1}{(x-z)^\lambda} \frac{d}{dz} \int_a^z \frac{f(\xi)}{(z-\xi)^{1-\lambda}} d\xi\, dz$$

$$= \int_a^x \frac{1}{(x-z)^\lambda} \frac{d}{dz} \int_a^z u(\xi) \int_\xi^z \frac{G(\xi_1, \xi)}{(z-\xi_1)^{1-\lambda}(\xi_1-\xi)^\lambda} d\xi_1\, d\xi\, dz \quad (7).$$

A reference to Theorem 3, § 1, shows that the first member of this equation is equal to

$$\frac{d}{dx} \int_a^x \frac{1}{(x-z)^\lambda} \int_a^z \frac{f(\xi)}{(z-\xi)^{1-\lambda}} d\xi\, dz.$$

By an application of Dirichlet's Extended Formula, this reduces to

$$\frac{d}{dx} \int_a^x f(\xi) \int_\xi^x \frac{dz}{(x-z)^\lambda (z-\xi)^{1-\lambda}} d\xi$$

$$= \frac{\pi}{\sin \lambda\pi} \frac{d}{dx} \int_a^x f(\xi)\, d\xi = \frac{\pi}{\sin \lambda\pi} f(x).$$

The second member of (7) may be written, as we see again by a reference to Theorem 3, § 1,

$$\frac{d}{dx} \int_a^x \frac{1}{(x-z)^\lambda} \int_a^z u(\xi) \int_\xi^z \frac{G(\xi_1, \xi)}{(z-\xi_1)^{1-\lambda}(\xi_1-\xi)^\lambda} \, d\xi_1 d\xi dz.$$

By a three-fold application of Dirichlet's Extended Formula this becomes

$$\frac{d}{dx} \int_a^x u(\xi) \int_\xi^x \frac{1}{(x-z)^\lambda} \int_\xi^z \frac{G(\xi_1, \xi)}{(z-\xi_1)^{1-\lambda}(\xi_1-\xi)^\lambda} \, d\xi_1 dz d\xi$$

$$= \frac{d}{dx} \int_a^x u(\xi) \int_\xi^x \frac{G(\xi_1, \xi)}{(\xi_1-\xi)^\lambda} \int_{\xi_1}^x \frac{dz}{(x-z)^\lambda (z-\xi_1)^{1-\lambda}} \, d\xi_1 d\xi$$

$$= \frac{\pi}{\sin \lambda \pi} \frac{d}{dx} \int_a^x \int_a^{\xi_1} u(\xi) \frac{G(\xi_1, \xi)}{(\xi_1-\xi)^\lambda} \, d\xi d\xi_1$$

$$= \frac{\pi}{\sin \lambda \pi} \int_a^x u(\xi) \frac{G(x, \xi)}{(x-\xi)^\lambda} \, d\xi.$$

Substituting the values we have just found on the two sides of (7), we see that this equation reduces to (1). We have thus proved the

THEOREM. *If $G(x, \xi)$ and its derivative*

$$G_1(x, \xi) = \partial G/\partial x$$

are continuous in T, and $G(x, x)$ does not vanish in I, a necessary and sufficient condition that the equation

$$f(x) = \int_a^x \frac{G(x, \xi)}{(x-\xi)^\lambda} u(\xi) \, d\xi \qquad\qquad (0 < \lambda < 1)$$

have a continuous solution is that $f(x)$ be continuous in I, that

$$f(a) = 0,$$

and that the function

$$F(x) = \int_a^x \frac{f(\xi)}{(x-\xi)^{1-\lambda}} \, d\xi$$

have a continuous derivative throughout I. If these conditions are fulfilled, the equation has only one continuous solution, namely the continuous solution of (4) where L is defined by (3).

By a reference to Theorem 3, § 1, it will be seen that a sufficient though not a necessary condition that F have a continuous derivative is that f be continuous and have a finite derivative with only a finite number of discontinuities.

If $G(x, \xi) \equiv 1$, equation (1) reduces to equation (1) of § 3. In this case

$$L = \pi/\sin \lambda \pi,$$

so that (4) becomes

$$\int_a^z \frac{f(x)}{(z-x)^{1-\lambda}}\, dx = \frac{\pi}{\sin \lambda\pi} \int_a^z u(\xi)\, d\xi,$$

from which, by differentiation, we get, under the assumptions there made, precisely the solutions (4) and (5) of § 3.

The theorem just proved may be readily modified so as to apply to cases in which

$$G(x, x) \equiv 0.$$

We leave it for the reader to carry through the discussion here.

A case which may be made to depend on the theorem of this section is that in which in equation (1), § 13, the function $K(x, x)$ vanishes identically while the derivative $K_1(x, y)$ has the form

$$K_1(x, y) = \frac{G(x, y)}{(x - y)^\lambda} \qquad (0 < \lambda < 1),$$

where G is continuous in T. Here too we leave the discussion and the precise formulation of the result to the reader. Theorem 4, § 3, is a special case of this.

An equation of some importance, which bears a certain resemblance to the equation of this section, is

$$f(x) = \int_0^1 \{ \operatorname{ctn} \pi (x - \xi) + G(x, \xi) \} u(\xi)\, d\xi,$$

treated by Hilbert and Kellogg. Here $G(x, \xi)$ is supposed to be finite in S, and, in order that the integral should have a meaning, it is in general necessary that we interpret it to mean the *principal value* according to Cauchy. For a treatment of this equation we refer to pages 17—27 of Kellogg's dissertation: "Zur Theorie der Integralgleichungen und des Dirichletschen Princips," Göttingen, 1902.

We close by a brief treatment of what was perhaps the first integral equation to be successfully treated.

One form of Fourier's Integral Theorem is contained in the formula

$$f(x) = \frac{2}{\pi} \int_0^\infty \int_0^\infty \cos(x\xi) \cos(\xi\xi_1) f(\xi_1)\, d\xi_1 d\xi \qquad (8).$$

This formula is valid when $0 \leqq x$ provided the function $f(x)$ satisfies certain conditions; for instance, it is sufficient to demand that, when $0 \leqq x$, $f(x)$ be continuous and do not have an infinite number of maxima and minima in the neighbourhood of any point, and that

$$\int_0^\infty |f(x)|\, dx$$

converge. Formula (8) shows us that, under the conditions just stated, the integral equation of the first kind

$$f(x) = \sqrt{\frac{2}{\pi}} \int_0^\infty \cos(x\xi)\, u(\xi)\, d\xi \qquad (9)$$

has as a continuous solution

$$u(x) = \sqrt{\frac{2}{\pi}} \int_0^\infty \cos(x\xi) f(\xi)\, d\xi \qquad (10).$$

Conversely, we see in the same way that (10) is the only continuous solution of (9) which can be expressed by means of Fourier's Integral Theorem.

The equation (9) has as its kernel the continuous symmetric function

$$\sqrt{\frac{2}{\pi}} \cos(x\xi) \qquad (11),$$

and the only peculiarity is that the interval I is now infinite. This change from a finite to an infinite interval makes a very essential difference in the whole theory of integral equations*, as we see by considering the characteristic numbers of the kernel (11). If the interval were finite, there would, as we know, be an infinite number of these numbers, each with a finite index†. Here, however, (at least if we restrict the conception of characteristic functions to functions for which Fourier's Integral Theorem (8) holds) there are only two characteristic numbers, namely ± 1, each with an infinite index. To prove this, let us consider the homogeneous equation of the second kind,

$$u(x) = \lambda \sqrt{\frac{2}{\pi}} \int_0^\infty \cos(x\xi)\, u(\xi)\, d\xi \qquad (12).$$

If λ is a characteristic number and $u(x)$ a corresponding characteristic function, not only will this equation (12) be fulfilled, but also, as we see by taking in (9) and (10) for $f(x)$ the function $u(x)/\lambda$,

$$u(x) = \frac{1}{\lambda} \sqrt{\frac{2}{\pi}} \int_0^\infty \cos(x\xi)\, u(\xi)\, d\xi \qquad (13).$$

---

* By introducing new variables

$$x' = \frac{x}{1+x}, \qquad \xi' = \frac{\xi}{1+\xi},$$

equation (9) with an infinite interval and a finite kernel may be reduced to an equation with a finite interval and an infinite kernel.

† We are here assuming the easily established fact that $\cos(x\xi)$ cannot be expressed as the sum of a finite number of terms each of which is the product of a function of x by a function of ξ.

By combining (12) and (13), we see that
$$u(x) \equiv \lambda^2 u(x),$$
so that, since $u(x)$ does not vanish identically, it follows that $\lambda = \pm 1$ are the only possible characteristic numbers.

To prove that these are really characteristic numbers and that they have an infinite index, we need merely notice that, when c has any positive value, the functions

$$\sqrt{\frac{\pi}{2}}\, e^{-cx} \pm \frac{c}{x^2 + c^2}$$

are solutions of (12) for $\lambda = +1$ and $\lambda = -1$ respectively.

The kernel (11) is only one of an important class of kernels having similar properties. For a theory of such kernels we refer to the dissertation by H. Weyl: "Singuläre Integralgleichungen mit besonderer Berücksichtigung des Fourierschen Integraltheorems," Göttingen, 1908.

INDEX

The numbers refer to pages

Printed in the United States
By Bookmasters